Volney Rattan

A Popular California Flora

Manual of botany for beginners. Containing descriptions of exogenous plants growing in central California, and westward to the ocean

Volney Rattan

A Popular California Flora

Manual of botany for beginners. Containing descriptions of exogenous plants growing in central California, and westward to the ocean

ISBN/EAN: 9783337316679

Printed in Europe, USA, Canada, Australia, Japan

Cover: Foto ©berggeist007 / pixelio.de

More available books at **www.hansebooks.com**

OR

MANUAL OF BOTANY FOR BEGINNERS.

CONTAINING

DESCRIPTIONS OF EXOGENOUS PLANTS GROWING IN CENTRAL CALIFORNIA, AND WESTWARD TO THE OCEAN.

By VOLNEY RATTAN,

TEACHER OF NATURAL SCIENCES IN THE GIRLS' HIGH SCHOOL,
SAN FRANCISCO.

SAN FRANCISCO:
A. L. BANCROFT AND COMPANY.
1879.

Entered according to Act of Congress, in the year 1879,
BY A. L. BANCROFT AND COMPANY,
In the Office of the Librarian of Congress, at Washington.

PREFACE.

This little book contains brief descriptions of over five hundred species of plants known to grow in the region bounded on the west by the coast line from Monterey Bay to Mendocino County, and on the east by the foothills of the Sierra Nevada. Only polypetalous and gamopetalous exogens are described, and of these the orders Umbelliferæ and Compositæ, with a few inconspicuous plants of other orders, are omitted, being considered too difficult for beginners.

The descriptions are mainly abridged from the large work on California Botany published in uniformity with the Geological Survey Reports. The Analytical Key is essentially the one prepared for that work by Dr. Asa Gray. Several names are changed and a few descriptions modified in accordance with "Gray's Synoptical Flora of North America."

Usually only contrasting characteristics of species are retained, since anything more is confusing rather than helpful to the beginner. The *habitat* of plants is given only when it would be of assistance in determining species.

All established common names are given. Fortunately our most beautiful plants are well known by their proper generic names.

A Second Part devoted to the Apetalæ and Edogenous plants, with an Introduction to Systematic Botany will probably be issued within a year.

Correspondents will confer a favor by pointing out errors and omissions.   V. R.

SAN FRANCISCO, February, 1879.

DIRECTIONS FOR THE LEARNER.

At first study only those plants which have large flowers. Do not attempt to determine the name of a plant unless you have specimens which show not only the flowers, but the buds, the fruit (at least, partly grown), the leaves from all parts of the stem, and the roots. If you can not readily distinguish the parts of the flower and their relations to each other, lay it aside until the study of easier plants has given you more skill.

If the small veins of the leaves do not form a network, and the organs of the flower are all in threes, the plant is probably an endogen and, therefore, not described in this book. If the calyx and corolla are not both present the flower is said to be apetalous. A few such flowers are described, and the orders are to be sought under the last division of the Analytical Key. *Clematis, Allotropa* and *Glaux* are apetalous genera having the calyx corolla-like.

It will be best to write out a description of the plant before attempting to analyze it. Some good text-book, such as Gray's "How Plants Grow," should be constantly referred to in search of the proper descriptive terms. Suppose the following to be a description of the plant in hand, the words in parentheses indicating a more concise way of telling the same thing: A hairy plant one or two feet high, with opposite leaves and no stipules (exstipulate); the leaves narrow and blunt, broader near the upper end (spatulate); the small pinkish flowers growing on short stems (pedicels) close together along one side of the main stem for several inches to the end (in a close raceme); the calyx of 5 sepals united to form a narrow tube (tubular, 5-lobed) nearly half an inch long, marked by 10 ridges (nerves); the 5 petals distinct from each other and very narrow inside the calyx (clawed), the part outside (blade) short and rounded, broadest near the end (obovate); two little teeth standing up on each petal just where it bends outward from the calyx (blade 2-appendaged at the base); the petals twisted so as to make one edge higher than the other; stamens 10; pistil one, with three short straight styles and a short stem below the ovary (stipe) on which the petals and stamens grow; the seed pod (capsule) ovoid and rough, containing seeds which grow fast to a central part (placenta).

Turning to the Analytical Key, we determine that it belongs under Division 1, because the petals are separate. Since our plant cannot belong under A, the stamens being only 10, we turn to "B. Stamens 10 or less," etc.; under this head we read: "1. *Ovary or ovaries superior,*" etc., which is the case with our plant; then follows: "* *Pistils more than one and distinct;*" but there is only one pistil in our flower, so we pass on to "* * *Pistil only one,*" below which is "+– *Simple, i. e., of one carpel, as shown by single style,*" etc., but there are three styles in the pistil of our flower, so we try " +– +– *Pistil compound,*" etc., which suits our case; then we read the next line, but upon looking up the word "placenta," conclude that our plant cannot be found under that head; the next line of the same length does not suit because our pod is not "2-celled;" but the third reads: "Ovary and capsule 1-celled, seeds on a central placenta," which applies to our seed pod; in the next line "Sepals 2; fleshy herbs" is wrong, so we try "Sepals 5 or 4; leaves opposite," etc., which leads us to ORDER CARYOPHYLLACEÆ, p. 27. The description of the order is satisfactory. Evidently our plant belongs to the first genus named in the Synopsis of Genera, viz.: *Silene;* and the first species under that genus is our plant, the proper name of which is *Silene Galica.* We find, too, as we always shall, that some things were omitted in our description; also, that all the characteristics of the plant are not given in this book. Having determined the name, you should next write out a description, as nearly complete as possible, and make drawings showing the outlines of the leaves and the separate parts of the flower.

ANALYTICAL KEY.

DIVISION 1. **Polypetalæ**; calyx and corolla both present, the latter of separate petals.
 A. Stamens more than 10, and more than double the number of petals.
  1. *Hypogynous, i. e., on the receptacle free from the other parts of the flower.*
  Pistils, few to many distinct carpels, rarely one.
   Calyx mostly deciduous; juice colorless .... Ranunculaceæ. 16
   Calyx early deciduous; juice yellowish .... Papaveraceæ, 20
   Calyx persistent; leaves peltate............................. Nymphæaceæ, 20
  Pistil compound; cells or stigmas more than one.
   Petals more numerous than the sepals.
    Indefinitely numerous, small and persistent; aquatic .... Nymphæaceæ, 20
    Just twice as many (4 or 6), and both usually caducous .... Papaveraceæ, 20
    Five to 16, and more numerous than the persistent sepals ..... Portulacaceæ, 29
   Petals of the same number as the sepals.
    Five, and the calyx persistent.
     Sepals valvate in the bud; stamens all united .... Malvaceæ, 31
     Sepals overlapping in the bud (imbricated).
      Leaves opposite, entire, transparent-dotted .... Hypericaceæ, 30
      Leaves alternate, not dotted (punctate).
       Two outer sepals smaller .... Cistaceæ, 25
  2. *Perigynous or epigynous (on the free or adnate calyx).*
  Fleshy herbs, with 3 or more cells to the ovary .... Ficoideæ, 63
  Fleshy herbs, with 1-celled ovary............................. Portulacaceæ, 29
  Not fleshy; herbs or shrubs.
   Leaves opposite, simple; sepals and petals numerous .... Calycanthaceæ, 55
   Leaves opposite, simple; sepals and petals 4 or 5. .... Saxifragaceæ, 55
   Leaves alternate, with stipules .... Rosaceæ, 49
   Leaves alternate, without stipules; rough herbs .... Loasaceæ, 62

 B. Stamens 10 or less, or if more, not exceeding twice the number of petals, or sepals.
  1. *Ovary or ovaries superior or mainly so (but sometimes inclosed in the calyx-tube).*
      * *Pistils more than one, and distinct.*
  Pistils of the same number as petals and sepals.
   Leaves simple, fleshy... Crassulaceæ, 58

ANALYTICAL KEY.

Leaves pinnate. (Styles partly united.)..........................Geraniaceæ, 33
Pistils not corresponding in number with the petals or sepals.
  Stamens borne on the receptacle..............................Ranunculaceæ, 16
  Stamens borne on the calyx.
    Stipules persistent; leaves alternate.............................Rosaceæ, 49
    Stipules none or indistinct...................................Saxifragaceæ, 55

* * *Pistil only one.*

+ *Simple, i. e., of one carpel, as shown by the single style, stigma, and cell (apparently 2-celled legume in Astragalus).*

Stamens opposite the petals and sepals, in threes; fruit a berry....Berberidaceæ, 19
Stamens 10, monodelphous or diadelphous; fruit a legume..........Leguminosæ, 38
Stamens on the calyx; fruit a drupe or akene........'................Rosaceæ, 49

+ + *Pistil compound, as shown by the cells or placentæ, styles or stigmas.*

Ovary 1-celled, placentæ (2 to 4, rarely more) on the sides.
  Petals and sepals 5; lower petal spurred............................Violaceæ, 25
  Petals and sepals 5; corolla regular.............................Saxifragaceæ, 55
  Petals 4, bract-like sepals 2; flower irregular...................Fumariaceæ, 22
  Petals 4 or 6; sepals half as many, caducous...................Papaveraceæ, 20
Ovary and pod 2-celled; stamens 6 (2 and 4).......................Cruciferæ, 22
Ovary and capsule 1-celled; seeds on a central placenta.
  Sepals 2; fleshy herbs..Portulacaceæ, 29
  Sepals 5 or 4; leaves opposite on swollen nodes...............Caryophyllaceæ, 27
  Petals and stamens 6; stems angled................................Lythraceæ, 59
  Sepals and petals (4 to 9) usually 5; shrubs; leaves 3-foliolate...Anacardiaceæ, 38
  Sepals 5; stamens 5, opposite the petals; herbs..............Plumbaginaceæ, 71
Ovary more than 1-celled; seeds attached to the axis or base or summit.
  Flowers very irregular; ovary 2-celled; pods flattened, 2-seeded..Polygalaceæ, 27
  Flowers regular or nearly so.
    Foliage pellucid-punctate; strong-scented shrubs.................Rutaceæ, 34
    Foliage not pellucid-punctate.
      Anthers opening by terminal pores..........................Ericaceæ, 68
      Anthers opening lengthwise.
        Stamens as many as the petals and opposite them.........Rhamnaceæ, 35
        Stamens alternate with the petals when equal in number.
          Leaves lobed or compound; stamens 10................Geraniaceæ, 33
          Leaves simple and entire; stamens 5....................Linaceæ, 32
          Leaves opposite and stipulate ....Geraniaceæ, 33
          Stamens on the calyx; style 1........................Lythraceæ, 59

ANALYTICAL KEY.

Stamens on the calyx; styles 2 or 3 .... Saxifragaceæ, 55
Shrubs or trees with opposite leaves.
   Pinnately veined, not lobed .... Celastraceæ, 35
   Palmately compound; stamens 4 to 8 .... Sapindaceæ, 37

     2. *Ovary and fruit inferior or mainly so.*

Tendril-bearing herbs; flowers monœcious, axillary .... Cucurbitaceæ, 63
Shrubs with opposite leaves; flowers diœcious .... Cornaceæ, 63
Shrubs; stamens opposite the clawed petals .... Rhamnaceæ, 35
Shrubs; stamens alternate with the petals.
   Leaves alternate; fruit a berry .... Saxifragaceæ, 55
   Leaves opposite, entire; style 1 .... Cornaceæ, 63
Herbs; fruit a many-seeded capsule .... Saxifragaceæ, 55
Herbs; parts of the flower mostly in fours (rarely 2 or 6) .... Onagraceæ, 59
Herbs; flowers in umbels; styles 2; fruit dry .... Umbelliferæ, 63

DIVISION 2. **Gamopetalæ**: petals more or less united into one piece.
A. Ovary inferior, or largely so.
Stamens more numerous than the lobes of the corolla, 8 or 10.
   Distinct and free from it, or nearly so .... Ericaceæ, 68
Stamens as many as the lobes of the corolla (5, rarely 4), united into a tube.
   Flowers in an involucrate head .... Compositæ, 66
   Flowers separate in racemes or spikes .... Lobeliaceæ, 67
Stamens as many as the corolla-lobes, distinct.
   Nearly or quite free; leaves alternate, no stipules .... Campanulaceæ, 67
   Inserted on the corolla; leaves opposite or whorled.
     With stipules, or in whorls, entire .... Rubiaceæ, 65
     Without stipules, opposite .... Caprifoliaceæ, 64
Stamens only 3, fewer than the lobes of the corolla.
   Leaves opposite; stamens distinct .... Valerianaceæ, 66
   Leaves alternate; stamens united .... Cucurbitaceæ, 63
B. Ovary superior (free), or mainly so.
     1. *Stamens more numerous than the lobes of the corolla.*
Pistil single and simple; leaves compound .... Leguminosæ, 38
Pistils several and simple; leaves simple, fleshy .... Crassulaceæ, 58
Pistil compound, with one undivided style .... Ericaceæ, 68
     2. *Stamens as many as the divisions of the corolla and opposite them.*
Styles 5; ovary 1-ovuled, fruit 1-seeded .... Plumbaginaceæ, 71
Styles 1; capsule at least several-seeded .... Primulaceæ, 72

3. *Stamens as many as the lobes of the corolla and alternate with them, or fewer.*
　　　* *No green herbage.*
Corolla regular; stamens free; seeds numerous, minute............Monotropeæ, 69
Corolla regular; stamens on the tube; twining parasites..........Convolvulaceæ, 86
Corolla irregular; stamens 4 in pairs...........................Orobanchaceæ, 96
　　　* * *With ordinary green herbage.*
　　　+ *Corolla regular or nearly so; stamens not in pairs.*
Corolla transparent and veinless; leaves all radical.............Plantaginaceæ, 103
Corolla more or less veiny.
　Stamens 5 or 4, as many as the corolla lobes.
　　Pollen in waxy masses; fruit a pair of follicles..............Asclepiadaceæ, 73
　　Pollen in powdery grains.
　　　Ovaries 2; fruit a pair of follicles......................Apocynaceæ, 73
　　　Ovary 4-lobed, forming 4 seed-like nutlets................Borraginaceæ, 83
　　　Ovary single and entire
　　　　Style 3-cleft; capsule 3-celled; corolla convolute........Polemoniaceæ, 75
　　　　Styles or stigmas 2 or 1.
　　　　　Seeds 4 at most, large; peduncles axillary............Convolvulaceæ, 86
　　　　　Seeds few or many; embryo small, in albumen.
　　　　　　Leaves opposite or whorled, entire; or
　　　　　　Leaves alternate, 3-foliolate, leaflets entire............Gentianaceæ, 74
　　　　　　Leaves various, mainly alternate.
　　　　　　　Styles 2, or 1 and 2-cleft; capsule 1-2-celled....Hydrophyllaceæ, 80
　　　　　　　Style 1; capsule or berry 2-celled ....Solanaceæ, 88
　　　　　　　See also Romanzoffia in......................Hydrophyllaceæ, 80
　　　　　　　And Limosella in...........................Scrophulariaceæ, 89
　　　+ + *Corolla irregular; stamens with anthers 4 in pairs, or 2; style 1.*
Ovary and capsule 2-celled.....................................Scrophulariaceæ, 89
Ovary 4-parted; fruit 4 seed-like nutlets.............................. Labiatæ, 97
Ovary undivided; fruit splitting into 2 or 4 nutlets................Verbenaceæ, 102

APETALOUS FORMS IN POLYPETALOUS AND GAMOPETALOUS ORDERS.

Carpels 1 or 2, rarely 3, distinct and free; stamens on the calyx.........Rosaceæ, 49
Capsule 2-celled, 2-seeded, flattened...................................Cruciferæ, 22
Shrub with alternate simple leaves; flowers 4-merous..............Rhamnaceæ, 35
Trees with opposite compound or lobed leaves.
　Fruit 1-seeded, winged (samara)......................................Oleaceæ, 73
　Fruit a pair of winged carpels (samaras).........................Sapindaceæ, 37
Herbs; ovary 1-celled; style and stigma 1; leaves opposite.........Primulaceæ, 72

GLOSSARY.

ABORTION, the imperfect formation or absence of a part.
ABRUPT, ending suddenly.
ACAULESCENT, apparently stemless.
ACCUMBENT, the radicle lying against the edges of the cotyledons.
ACEROSE, needle-shaped, like pine leaves.
ACUMINATE, ending in a tapering point.
ACUTE, merely sharp-pointed.
ADNATE, growing fast to. When the anther seems to be attached by its whole length to the filament.
AGGREGATE, crowded into a cluster.
AKENE, a 1-seeded seed-like fruit.
ALBUMEN, nourishment in the seed not forming part of the embryo.
ANDROUS, refers to stamens.
ANTERIOR, on the side of the flower next the bract.
APETALOUS, without petals.
APPRESSED, lying flat, or close together.
ASCENDING, rising obliquely.
ATTENUATE, tapering gradually.
AURICULATE, ear-like lobes at the base.
AWN, an appendage like the beard of barley.
AXIL, the angle between leaf and stem.

BIFID, 2-cleft to about the middle.
BILABIATE, 2-lipped.
BLADE, the broad portion of a leaf.

BRACT, the leaf which subtends the flower.
BRACTLET, a bract on a pedicel.

CADUCOUS, falling off at the time of expansion.
CAMPANULATE, bell-shaped.
CANESCENT, whitened with fine close pubescence.
CAPILLARY, like a hair.
CAPITATE, having a head, or collected into a head.
CAPSULE, any compound dehiscent fruit.
CARPEL, a simple pistil, or element of a compound one.
CAUDATE, tailed.
CAULESCENT, having an obvious stem.
CAULINE, relating to a stem.
CILIATE, fringed with hairs.
CLAVATE, club-shaped.
CLAW, the narrowed base of a petal.
CLEFT, cut to about the middle.
COHESION, the union of like organs.
CONFLUENT, running together, or blending.
CONGLOMERATE, thickly clustered.
CONNATE, united from the first.
CONNECTIVE, the part of an anther connecting the cells.
CONNIVENT, coming together or meeting.
CONVOLUTE, rolled up.
CORDATE, heart-shaped with the point up.

GLOSSARY.

CORYMB, a flat-topped flower cluster, the pedicels unequal.
COSTATE, ribbed.
COTYLEDONS, the leaves of the embryo.
CREEPING, running on the ground and rooting.
CRENATE, the margin scolloped.
CUNEATE, wedge-shaped.
CUSPIDATE, tipped with a rigid point.
CYME, a flower cluster in which the oldest flowers are in the center.

DECIDUOUS, falling off before withering; or, if leaves, before winter.
DECLINED, turned to one side.
DECUMBENT, reclining on the ground, the end rising.
DEFLEXED, bent downwards.
DEHISCENT FRUITS, etc., open by
DEHISCENCE, splitting as pods do.
DENTATE, toothed, the teeth pointing directly away from the margin.
DEPRESSED, flattened from above.
DIADELPHOUS, stamens united by the filaments in two sets.
DICHOTOMOUS, forking into two branches.
DICOTYLEDENOUS, having two seed leaves.
DIFFUSE, widely and loosely spreading.
DIGITATE, compound with the parts arising at one point.
DIŒCIOUS, with stamens and pistils in separate blossoms on different individuals.
DISSECTED, cut into pieces, or nearly so.
DISTINCT, when parts of the same name do not cohere.
DIVARICATE, separating widely.
DIVERGENT, the summits inclined from each other.
DRUPE, a stone fruit (like a cherry).

EMBRYO, the rudimentary plant in a seed.
ENTIRE, the margin whole and even, not lobed or toothed.
EPIGYNOUS, growing on the ovary.
EROSE, irregularly notched as if gnawed.
EXSERTED, protruding beyond other organs.
EXSTIPULATE, without stipules.
EXTRORSE, turned outward.

FASCICLE, a close cyme, a bundle of leaves.
FERTILE FLOWER, one having pistils.
FILAMENT, the stalk of an anther.
FILIFORM, like a thread.
FOLIACEOUS, like a leaf.
FOLIOLATE, consisting of leaflets (5-foliolate means with five leaflets).
FOLLICLE, a simple pod opening down one side.
FRUIT, the seed and all that belong to it.

GLAUCOUS, covered with a whitish bloom which rubs off, as the surface of a cabbage leaf, or a plum.
GLOMERATE, clustered into a ball.
GLOMERULE, a capitate cyme.

HASTATE, with a spreading lobe at the base on each side.
HIRSUTE, clothed with coarse hairs.
HISPID, beset with bristly hairs.
HOARY, grayish white from a white pubescence.
HYPOGYNOUS, growing under the pistil, free from the calyx and corolla.

INCUMBENT, when the radicle lies against the back of one of the cotyledons.
INFERIOR, underneath or anterior.
INNATE, borne on the apex or end.
INTRORSE, turned inward.

GLOSSARY. 13

INVOLUCRE, a set of bracts surrounding a flower cluster.
INVOLUTE, rolled inward.
IRREGULAR, unequal in size or shape.

LACINIATE, cut into narrow incisions.
LAMINA, blade of a leaf or petal.
LATERAL, pertaining to the side.
LEGUME, fruit like a pea-pod.
LIMB, the exposed part of a corolla, calyx, etc., or the blade of a petal, etc.
LINE, the twelfth of an inch.
LINEAR, narrow and much longer than wide, the margins parallel.
LOBE, any division or projecting part.

MEROUS, the parts of a flower (5-merous, the parts in fives).
MUCRONATE, abruptly tipped with a short point.

NERVES, parallel and simple veins.
NODDING, the apex or top pointing downward.

OB-, prefixed means reverse of; as, ob-cordate, inverted heart-shaped, *i. e.*, the stem attached to the apex.
OBLIQUE, one-sided.
OBLONG, long-elliptical.
OCHROLEUCOUS, pale dull yellow.
OVAL, broadly elliptical.
OVARY, that portion of the pistil which becomes the seed vessel.
OVATE, like the longitudinal section of an egg.
OVOID, egg-shaped.

PALMATE, lobed so that the lobes point away from the end of the petiole, as in an ivy or a maple leaf.

PANICLE, a raceme branching irregularly.
PARTED, cut almost through.
PECTINATE, like the teeth of a comb.
PEDICEL, the stalk of a single blossom in a cluster.
PEDUNCLE, the stalk of a cluster or of a solitary flower.
PERFOLIATE, when the stem seems to pass through the leaf.
PERFORATE, with holes or transparent dots.
PERIGYNOUS, borne on the calyx.
PERSISTENT, remaining until the fruit has grown.
PETIOLE, the leaf stem.
PETIOLULE, the stem of a leaflet.
PILOSE, with distinct straight hairs.
PINNATE, a compound leaf with the leaflets along the side of a common petiole.
PINNATELY CLEFT, LOBED, etc., with the lobes along the sides of a long leaf.
PLACENTA, the part of the ovary which bears the seeds.
POD, a dry dehiscent fruit.
POME, a fruit like a pear or apple.
POSTERIOR, next the stem.
PROCUMBENT, lying along the ground.
PROSTRATE, lying flat like a melon-vine.
PUBESCENT, with soft or downy hairs.
PUNCTATE, dotted as if by holes.
PUNGENT, rigid sharp-pointed.

RACEME, elongated flower bunches, with the oldest flowers below and on pedicels.
RADICAL, coming from the root (apparently).
RADICLE, the stem of an embryo.
RENIFORM, kidney-shaped.
REPAND, the margin slightly wavy.
RETRORSE, directed backward.
RETUSE, slightly notched at a rounded apex.

REVOLUTE, rolled backward.
RACHIS, the main stem in a spike, etc.
ROOTSTOCK, an underground stem.
ROTATE, wheel-shaped.
RUNCINATE, teeth pointing backward.

SAGITTATE, like an arrow-head.
SALVER-SHAPED, tubular, the border spreading at right angles to the tube.
SCAPE, a flower-stalk rising from the ground or near it.
SCORPIOID, coiled round like a scorpion.
SECUND, all turned to one side.
SERRATE, with teeth like a saw.
SETACEOUS, like a bristle.
SPATULATE, like a druggist's spatula.
SPIKE, a long inflorescence of sessile flowers.
STELLATE, star-shaped.
STIGMA, the part of a pistil which receives the pollen.
STIPE, the stalk of an ovary.
STIPEL, the stipule of a leaflet.
STIPELLATE, having stipels.
STIPITATE, having a stipe.

STIPULE, appendage on each side at the base of a leaf.
STRICT, very straight or close or upright.
STRIGOSE, clothed with close-pressed stout sharp hairs or scale-like bristles.
STYLE, the slender part of a pistil.
SUBULATE, tapering to a sharp rigid point.
SUFFRUTESCENT, or *suffruticose*, shrubby at the base.

TERETE, cylindrical, long and round.
TERMINAL, at the end or summit.
THYRSE, a thick panicle (Lilac blossoms).
TOMENTOSE, clothed with a close and matted down.
TORULOSE, swollen at intervals.
TRUNCATE, as if cut off at the end.

UMBEL, umbrella-like inflorescence.

VERTICILLATE, whorled, forming a ring around the stem.
VILLOUS, with long soft hairs.
VISCID, sticky.

BOTANY

OF

WEST-CENTRAL CALIFORNIA.

SERIES I.

FLOWERING OR PHÆNOGAMOUS PLANTS.

Plants producing flowers and seeds; the former consisting, at least, of stamens and pistils, which may be together in the same flower, or they may separately form staminate and pistillate flowers growing on the same individual, or different individuals of one species; the latter containing a germ, or embyro.

CLASS I.—EXOGENOUS DICOTYLEDONS.

Stems consisting of pith in the center, bark on the outside, and between these, fibrous or woody tissue, which, in perennial stems, increases from year to year by the addition of layers on the outside next the bark. Embryo usually of two opposite cotyledons, or rarely with several in a whorl.

SUB-CLASS I.—ANGIOSPERMS.

Pistil consisting of a closed ovary which forms the fruit. Cotyledons two.

RANUNCULACEÆ. (CROWFOOT FAMILY.)

DIVISION I. POLYPETALÆ.

Order 1. RANUNCULACEÆ.

Herbs or shrubs, with colorless juice; foliage various; stipules none; organs of the flower free and distinct; sepals, petals, and pistils few or many; stamens numerous; petals sometimes wanting, then the sepals are usually petaloid; anthers short and adnate; seeds with minute embryos in fleshy albumen.

Flowers regular.

Petals none; shrubby climbers..Clematis. 1
Petals none; small herbs..Anemone. 2
Petals 5 or more; carpels numerous.................................Ranunculus. 3
Petals 5, spurred; carpels 5..Aquilegia. 4

** *Flowers irregular; colored sepals conspicuous.*
Upper sepal spurred..Delphinium. 5
Upper sepal hooded...Aconitum. 6

*** *Sepals large, leaf-like, persistent.*
Flowers large..Pæonia. 7

1. CLEMATIS, L.   Virgin's Bower.

Sepals 4, colored and petal-like, valvate in the bud. Pistils numerous; styles persistent, becoming long feathery tails in fruit. Half-woody climbers or perennial herbs, with opposite leaves.

1. C. ligusticifolia, Nutt. Stems climbing by the petioles of the 5-foliolate leaves; leaflets broadly ovate to lanceolate, 1½ to 3 inches long, acute or acuminate, 3-lobed and coarsely toothed, rarely entire or 3-parted. Flowers diœcious, paniculate; sepals thin, silky, white, 4 to 6 lines long; akenes pubescent; tails 1 to 2 inches long.

Var. Californica, Watson. Leaves silky-tomentose beneath, often small.

2. C. lasiantha, Nutt. Leaves 3-foliolate; leaflets ovate, 1 to 1½ inches long, acute, coarsely toothed or 3-lobed or the terminal 3-parted. Flowers solitary on 1-2-bracted peduncles; sepals obtuse, thick, 6 to 10 lines long.

2. ANEMONE, L.

Sepals 4 to 20, colored and petal-like, imbricated in the bud. Petals none. Pistils numerous; style short; stigma lateral; akenes compressed, pointed, in a head. Erect perennial herbs, with lobed or divided leaves, which are radical, except those which form an involucre below the flower.

RANUNCULACEÆ. (CROWFOOT FAMILY.) 17

1. **A. nemorosa**, L. (WOOD ANEMONE.) Smooth or somewhat villous; stems from a slender rootstock, 3 to 12 inches high, without radical leaves, one-flowered; involucre of 3 petioled ternate leaves, the divisions cuneate-oblong to ovate, incisely toothed or lobed, or the lateral ones 2-parted, about an inch long; the 4 to 7 sepals pinkish or white; akenes 12 to 20, oblong, with a hooked beak.

Here belongs *Thalictrum Fendleri*, Englm. A smooth apetalous diœcious herb; also, *Myosurus minimus*, L. A very small herb, with a tuft of linear or spatulate entire radical leaves, and solitary flowers on simple scapes; called *Mouse-tail*, from its long, narrow receptacle, densely covered with small akenes.

3. RANUNCULUS, L. BUTTERCUP.

Sepals usually 5. Petals 3 to 18. Pistils numerous. Akenes in a head, usually flattened, beaked with the persistent style.

§ 1. *Aquatic herbs; petals white, with a pit at the base, the claw yellow; akenes transversely wrinkled.*

1. **R. hederaceus**, L., var. Glabrous; stems 6 to 12 inches long, floating; leaves commonly all floating, 3 to 8 lines wide, deeply 3-lobed, truncate or cordate at the base; the lobes equal, oval or oblong, the lateral ones usually with a broad notch in the apex; submersed leaves none or rudimentary and resembling adventitious roots; peduncles opposite the upper leaves, thicker than the petiole, 6 to 8 lines long; sepals a line long; petals 2 lines long, obovate oblong; stamens 5 to 9; akenes 4 to 6.

2. **R. aquatilis**, L., var. **tricophyllus**, Chaix. Stems long, filiform; leaves all submersed and cut into numerous capillary segments, which are 4 to 10 lines long; flowers 3 to 5 lines in diameter; akenes numerous in a globular head.

§ 2. *Terrestrial herbs, but often growing in wet places; sepals green; petals yellow, with a scale at the base; akenes neither wrinkled nor hispid.*

* *All the leaves undivided, the margins entire.*

3. **R. Flammula**, L., var. **reptans**, Gr. Glabrous throughout; stems filiform, creeping and rooting at the joints, 4 to 10 inches long; leaves mostly lanceolate and acute at each end, entire; flowers 2 to 5 lines in diameter; petals broadly obovate, one half longer than the sepals; akenes few, in a small globular head, plump, smooth; beak very short and curved.

4. **R. alismæfolius**, Geyer. Similar to the last species, but with stoutish, erect stems, longer flowers and obtuse leaves; akenes straight-beaked.

* * *Some or all the leaves ternately compound.*

5. **R. Californicus**, Benth. More or less hairy; stems erect, or nearly so, 12 to 18 inches high; radical leaves, commonly pinnately ternate, the leaves laciniately cut into 3 to 7 parts, which are usually linear; flowers bright yellow, 5 to 10 lines in diameter;

18 RANUNCULACEÆ. (CROWFOOT FAMILY.)

petals 10 to 14, narrowly obovate; sepals shorter than the petals, reflexed; akenes nearly 2 lines long, flat, with sharp edges; beak short and curved; heads compact, ovate or globular.
This is by far the most common species, and usually the only one collected by beginners. It varies greatly. The leaves are sometimes simply three lobed and sometimes much cut up.
6. R. macranthus, Scheele. Stems stout, 2 to 5 ft. high; flowers 14 to 18 lines in diameter; petals commonly 5 or 6, broadly obovate, shining yellow.

§ 3. *Akenes rough; otherwise as in* § 2.

7. R. hebecarpus, Hook. & Arn. Rather slender, more or less hairy; flowers minute; petals 5, not more than a line long; sepals hairy, about equaling the petals.
8. R. muricatus, L. Smooth; flowers 5 or more lines in diameter; akenes large and rough, with recurved beaks. Introduced from Europe.

4. AQUILEGIA, Tourn. COLUMBINE.

Sepals 5, regular, colored and petal-like; petals 5, produced backward (upward) into a long tubular spur; stamens numerous, exserted, the inner ones reduced to thin scales; pistils 5; styles slender. Flowers nodding, showy, terminating the branches.
1. A. truncata, Fisch. & Mey. Stems 1 to 3 ft. high; flowers usually red, tinged with orange or yellow; leaves usually ternately compound, leaflets lobed.

5. DELPHINIUM, Tourn. LARKSPUR.

Sepals 5, colored and petal-like, very irregular, the upper one prolonged backwards at the base into a long spur, which (in our species) contains spur-like prolongations of the upper pair of petals. Petals 4, small and irregular. Stamens many. Pistils 1 to 5. Erect herbs, with palmately-cleft, lobed, or dissected leaves, and racemose flowers.
1. D. simplex, Dougl. Canescent throughout, with a fine, short, somewhat woolly pubescence, rarely smooth; stem stout and strict, 1 to 3 ft. high, leafy; leaves all much dissected with linear obtuse lobes, on stout, erect petioles; racemes usually dense and many-flowered, the pedicels often short and nearly erect; flowers small, blue, varying to nearly white or yellowish; sepals 4 or 5 lines long, about equaling the stout, straight spur; ovaries and capsule pubescent.
2. D. variegatum, Torr. & Gr. Foliage similar to the last, but the flowers much larger, on longer pedicels, forming a short, open raceme; ovary and capsule pubescent.
3. D. decorum. Fisch. & May. Lower leaves 5-lobed, sparingly toothed, the upper with narrow divisions. Flowers similar to the last, but the spur is usually longer, and the ovary and capsule smooth.
4. D. Californicum, Torr. & Gr. Stems stout, 2 to 7 ft. high; leaves large, 3 to

5 cleft, the divisions variously lobed; pedicels and dull bluish flowers densely velvety pubescent.
5. **D. nudicaule**, Torr. & Gr.   Distinguished by its red flowers.

6. ACONITUM, Tourn. MONKSHOOD.

Sepals 5, colored and petal-like, very irregular; the upper one arched into a hood or helmet, which conceals the spur-like blades of the upper pair of petals. General appearance similar to *Delphinium*.
1. **A. Fischeri**, Reichenb.   Sufficiently characterized by the generic description. Rare.

7. PÆONIA, L.

Sepals 5, herbaceous. Petals 5 to 10. Stamens inserted on a fleshy disk. Pistils 2 to 5. Fruit leathery follicles. Perennial herbs with compound leaves.
1. **P. Brownii**, Dougl.   Leaves thick, 1-2-ternately compound, the leaflets ternately and pinnately lobed, glaucous; petals leathery, dull, dark red, about equaling the sepals.

ORDER 2. BERBERIDACEÆ.

Shrubs or herbs, with compound alternate exstipulate leaves; flowers remarkable for having the bracts, sepals, petals and stamens before each other, instead of alternating.
Low shrubs, with rigid pinnate leaves and small yellow flowers..........Berberis. 1
A fern-like herb, with white flowers.............................Vancouveria. 2

1. BERBERIS, L. BARBERRY.

Sepals, petals, and stamens 6 each, with 3 or 6 bractlets. Carpel 1, forming a berry. Smooth shrubs, with yellow wood, and yellow flowers in bracteate racemes.

* Leaflets pinnately veined.

1. **B. repens**, Lindl.   Less than a foot high; leaflets 3 to 7, ovate, acute, 1 to 2½ inches long, not shiny above; short racemes terminating the stems.
2. **B. Aquifolium**, Pursh.   2 to 4 ft. high; leaflets 7 or more, the lower pair distant from the stem, 1½ to 4 inches long, shining above, spiny; racemes chiefly clustered in subterminal axils.
3. **B. pinnata**, Lag.   Like the last species, but the leaves more crowded, and the lower pair of leaflets near the base of the petiole; usually 5 to 7 leaflets.

** Leaflets palmately nerved.

4. **B. nervosa**, Pursh.   Simple stems but a few inches high; leaves 1 to 2 ft. long, of 11 to 17 leaflets.

PAPAVERACEÆ. (POPPY FAMILY.)

2. **VANCOUVERIA**, Morren & Decaisne.

Sepals and petals 6 each, reflexed, with 6 to 9 bractlets. Stamens 6. Carpel 1; the stigma cup-shaped. A slender perennial herb, with radical 2-3-ternately compound leaves, and the open paniculate raceme upon a naked scape.

1. **V. hexandra**, Morr. & Dec. The long petioled leaves rising like the fronds of a fern, leaflets 1 to 2 inches broad, petiolulate, obtusely 3-lobed, the margin thickened; the minute flowers on a scape exceeding the leaves.

ORDER 3. **NYMPHÆACEÆ**.

Aquatic perennial herbs, with peltate or deeply cordate leaves; solitary axillary perfect flowers on long peduncles. Stamens numerous.

Water-Shield. (*Brasenia peltata*, Pursh.) May be found in ponds. Its elliptical, peltate, floating leaves (green above and brownish-red beneath) and its jelly-coated stems characterize it quite well enough.

The *Yellow Pond-Lily* (*Nuphar polysepalum*, Engl.) is more common.

ORDER 4. **PAPAVERACEÆ**.

Herbaceous plants (in one instance shrubby); the perfect flowers with sepals, petals and stamens hypogynous and not in fives; the sepals 2 or 3, and falling when the flower opens; the petals twice as many, in two sets; the stamens indefinite. In *Eschscholtzia* the sepals unite to form a miterform cap.

* *Herbs with entire leaves, the uppermost whorled or opposite.*
Carpels 6 to 25, slightly cohering..................................**Platystemon.** 1
Carpels 3 united; 3 stigmas...**Platystigma.** 2

** *Herbs or shrubs with divided or lobed leaves.*
Sepals 2...**Meconopsis.** 3
Sepals united to form a cap..**Eschscholtzia.** 4
Shrub, with entire leaves..**Dendromecon.** 5

1. **PLATYSTEMON**, Benth. CREAM-CUPS.

Sepals 3. Petals 6. Stamens many, with flattened filaments and linear anthers. Torulose carpels at first united; stigmas free.

1. **P. Californicus**, Benth. Slender, branching, 6 to 12 inches high; villous, with spreading hairs; leaves 2 to 4 inches long, sessile or clasping, broadly linear, obtuse,

PAPAVERACEÆ. (POPPY FAMILY.)

pale-green. Sepals hairy; petals pale-yellow, shading to orange in the center, 3 to 6 lines long.

2. PLATYSTIGMA, Benth.

Sepals 3. Petals 4 to 6. Stamens few or many, with narrow filaments. Ovary 3-angled, oblong or linear; stigmas 3, ovate to linear. Low, slender annuals, resembling *Platystemon* in habit, with pale-green, entire, opposite or verticillate leaves and long-peduncled pale-yellow or creamy-white flowers.

1. **P. lineare**, Benth. Hairy, short-stemmed; stamens many, with dilated filaments; stigmas broad; capsule ovate.

2. **P. Californicum**, Benth. & Hook. Smooth, long-stemmed; stamens few (10 to 12) with filiform filaments; stigmas narrow; capsule linear.

3. MECONOPSIS, Viguier.

Sepals 2. Petals 4. Stamens numerous, with filiform filaments and oblong anthers. Style distinct; stigma 4–8-lobed. Seeds numerous.

1. **M. heterophylla**, Benth. Annual, smooth, slender, 1 to 2 ft. high; lower leaves long petioled, pinnately divided, the segments oval to linear and 2 to 12 lines long; upper leaves sessile; flowers scarlet to orange, the petals 2 to 12 lines long; peduncles elongated. Very variable.

4. ESCHSCHOLTZIA, Chamisso.

Sepals coherent into a narrow pointed hood, which drops off from the top shaped torus when the flower opens. Petals 4. Stamens numerous, with short filaments and long anthers. Smooth annuals, with colorless, bitter juice; finely dissected, pale-green alternate petioled leaves, and bright orange or yellow (rarely white) flowers.

1. **E. Californica**, Cham. Has stout branching stems, 1 to 1½ ft. high; flowers 2 to 4 inches in diameter, brilliant orange toward the center; capsule 2½ inches long, curved.

Var. **Douglasii**, Gr. More slender; flowers yellow.

Var. **cæspitosa**, Brewer. Scape-like peduncles; small yellow flowers.

5. DENDROMECON, Benth.

Sepals 2. Petals 4. Stamens numerous, with short filaments and linear anthers. Ovary linear; style short; stigmas 2, short and erect. The many seeded capsule dehiscent the whole length by 2 valves separating from the placental ribs. A smooth branching shrub, with alternate vertical entire thick and rigid leaves and showy yellow flowers. The only true woody plant belonging to the order.

1. **D. rigidum**, Benth. A shrub 2 to 8 ft. high, with slender branches and whitish bark; leaves ovate to linear-lanceolate, 1 to 3 inches long, very acute or mucronate, sessile or nearly so, twisted into a vertical position, margin rough or denticulate.

22   CRUCIFERÆ. (MUSTARD FAMILY.)

Order 5. FUMARIACEÆ.

Tender herbs with dissected compound leaves, and irregular hypogynous flowers, the parts in twos, except the 6 diadelphous stamens.

1. DICENTRA, Borkh.

Sepals 2, small and scale-like, sometimes caducous. Corolla of two pairs of petals, flattened and cordate; the outer pair the larger and sacked at the base, the tips spreading; the inner, spoon-shaped, lightly united at the apex, inclosing the anthers and stigma. Stamens in two sets, 3 before each of the outer petals, filaments slightly cohering. Style slender; stigma 2-lobed, each lobe sometimes 2-crested.

1. **D. formosa**, DC.   Leaves radical, and the compound racemes of rose-colored flowers borne on naked scapes.

2. **D. chrysantha**, Hook. & Arn. The flowers in long terminal paniculate racemes on leafy stems; corolla narrow, scarcely cordate, golden yellow.

Order 6. CRUCIFERÆ.

Herbs with pungent watery juice. Sepals 4. Petals 4, with blade narrowed into a claw, the lamina spreading to form a cross, rarely wanting. Stamens 6, two of them inserted lower down on the receptacle and shorter than the other four. Ovary 2-celled by a thin partition, rarely 1-celled. Leaves alternate, and flowers usually in racemes without bracts.

Since a careful examination of the fruit is usually necessary for the determination of species in this difficult order, only such plants as have large flowers or remarkable fruit are here described.

§ 1. *Pod dehiscent, 2-valved.*

* *Pod elongated, compressed parallel with the partition; seeds flat.*

Petioled leaves, lobed or divided; root tuberous......................Cardamine. 1
Stem leaves sessile, entire; root perpendicular.
Flowers purple...Arabis. 2
Flowers orange...Cheiranthus. 3
Flowers yellowish..Erysimum. 4

* * *Pod terete; seeds globose.*

Flowers yellow...Brassica. 5

* * * *Pod flattened contrary to the partition.*

Pod linear; flowers axillary, yellow...........................Tropidocarpum. 6
Pod obcordate; flowers minute......................................Capsella. 7
Pod obovate, 2-winged at the top..................................Lepidium. 8

CRUCIFERÆ. (MUSTARD FAMILY.)

§ 2. *Pod indehiscent, 1-celled.*

Pod orbicular, winged with a thin broad margin; flowers minute...**Thysanocarpus.** 9
Pod long, pithy; seeds large; flowers large, veiny.................**Raphanus.** 10

1. CARDAMINE, L.

Pod linear, with somewhat thickened margins, merely pointed or beaked above; valve flat, nerveless. Seeds in one row somewhat flattened, wingless; cotyledons flat, accumbent. Sepals equal. Petals white or pinkish.

1. **C. paucisecta,** Benth. Stems from small deep-seated tubers, erect, 8 to 18 inches high; leaves various; the upper deeply lobed or parted, the lower often simple; petals 6 to 9 lines long; pods 1 to 1½ inches long.

2. ARABIS, L.

Pod linear; valves 1-nerved, not strongly. Seeds in 1 or 2 rows, flattened; cotyledons accumbent. Sepals short or narrow, rarely colored. Petals with a narrow claw, white, rose-colored, or purple.

1. **A. blepharophylla,** Hook. & Arn. Stems often tufted 4 to 12 inches high; leaves strongly ciliate, sometimes sparingly sinuate-toothed, the lower obovate or broadly spatulate, the cauline oblong, sessile; petals bright purple, 6 to 9 lines long.

2. **A. Breweri,** Wat. Cespitose, canescent, with dense stellate pubescence; stems 2 to 10 inches high; petals 1 to 4 lines long, deep rose-color; sepals purplish; pods spreading or recurved.

3. CHEIRANTHUS, L.

Pod elongated, compressed; valves 1-nerved or carinate. Seeds in one row, flattened, not winged; cotyledons accumbent, or rarely oblique. Calyx not colored, the outer sepals strongly gibbous. Stigma with two spreading lobes.

1. **C. Asper,** Cham. & Sch. Rather sparingly pubescent with appressed 2-parted hairs; stem simple erect, leafy, 1 to 3 ft. high; leaves spatulate or oblanceolate, the lower long petioled, entire or sinuate-toothed; sepals broad 4 to 6 lines long, half the length of the bright yellow or orange petals; pods 1½ to 2 inches long.

4. ERYSIMUM, L.

Pod 4-angled by the prominent mid-nerve of the valves, not stipitate; cotyledons incumbent or oblique. Sepals, petals and stigma like the last.

1. **E. asperum,** DC. Similar to the last; sepals narrower; petals usually creamy white to yellow.

5. BRASSICA, L. MUSTARD.

Pod nearly terete or somewhat 4-sided, pointed with a long conical beak. Seeds in one row globose; cotyledons infolding the radical. Lateral sepals usually gibbous. Petals yellow.

1. **B. campestris**, L. Smooth; lower leaves pinnately divided, with a large terminal lobe; the upper leaves oblong or lanceolate, with a broad clasping base; pods 2 inches long or more.

2. **B. nigra**, Boiss. Larger; leaves all petioled; pods less than an inch long.

Not to be confounded with **Sisymbrium officinale**, Scop., which has runcinately pinnatifid leaves, small yellow flowers and closely appressed, subulate sessile pods half an inch long; or, with **S. acutangulum**, DC., similar to the last, but the pods on short pedicels, erect and over an inch long. The last are called *Hedge Mustards*.

6. TROPIDOCARPUM, Hook

Pod linear, flattened, often 1-celled by the disappearance of the narrow partition. Seeds in two rows, minute; cotyledons incumbent. A low hirsute branching annual, with pinnately divided leaves, and yellow, solitary axillary flowers.

1. **T. gracile**, Hook. Stems weak; petals 1½ to 3 lines long, broad; pods 6 to 20 lines long, pointed at both ends.

7. CAPSELLA, Mœnch. SHEPHERD'S PURSE.

Pod obcordate, much flattened, many-seeded; cotyledons incumbent. Slender and mostly smooth annuals, with minute flowers.

1. **C. Bursa-pastoris**, Mœnch. Somewhat hirsute at base; radical leaves mostly runcinate-pinnatifid, the cauline lanceolate, clasping.

2. **C. divaricata**, Walp. Very slender; pods elliptic-oblong; is more rare.

8. LEPIDIUM, L. PEPPERGRASS.

Pod orbicular or obovate, emarginately 2-winged at the summit; the cells 1-seeded. Low herbs, with pinnatifid or toothed leaves, and small white flowers; the petals in some species wanting, and the stamens only 2 or 4.

1. **L. latipes**, Hook. Stems stout, simple 1 to 3 inches high, surpassed by the irregularly and coarsely pinnatifid leaves; racemes capitate, in fruit an inch long or less; sepals very unequal; pod strongly reticulated, the acute wings nearly as long.

2. **L. oxycarpum**, Torr. & Gr. Stems simple or branched 3 to 6 inches high; smooth; raceme lax, elongated; pod smooth, rounded, nodding, the broad acute teeth short and divergent; petals none.

3. **L. nitidum**, Nutt. Similar to the last, but larger; petals present; pods smooth and shining, acutely margined.

4. **L. Menziesii**, DC. Hispid; petals none; pods not margined, except by the very short teeth at the summit.

Var. (?) **strictum**, Wat. Sepals green, persistent; fruiting racemes crowded cylindric-capitate, the pedicels erect, low and spreading. This plant seems to be a separate species. It has been found in San Francisco, by *Miss Annie Hughes*.

9. THYSANOCARPUS, Hook.

Pod 1-celled, 1-seeded, plano-convex, mostly pendulous on slender pedicels. Flowers minute, white or rose-colored.

1. **T. curvipes**, Hook. Six inches to two feet high; the upper leaves clasping by a broad auricled base; pods densely tomentose or smooth, 2 to 4 lines in diameter, the wing entire or crenate, veined and often perforate, emarginate at the top and tipped with the purple style. The perforate-wing form called *Lace-pod*.
2. **T. laciniatus**, Nutt. Smaller and more slender; the cauline leaves scarcely auricled at the base; pods obovate, cuneate at the base, 2 to 3 lines long.
Var. **crenatus**, Brewer. The broader wing deeply crenate or fringed. *Fringe-pod*.
3. **T. radians**, Benth. Pods round, 4 to 5 lines in diameter, scarcely emarginate, with a broad entire translucent wing conspicuously marked by radiating nerves.
4. **T. pusillus**, Hook. May be known by its minute pods hirsute with hooked hairs.

10. RAPHANUS, L. Radish.

Coarse introduced annuals.
1. **R. sativus**, L., has a pointed 2-seeded pod.
2. **R. Raphinistrum**, L., has a necklace-shaped pod, long beaked, 1-9-seeded.

Order 7. CISTACEÆ.

Flowers perfect and regular. Sepals 5, persistent; and two of them smaller, wholly exterior, and bract-like. Petals 5, usually ephemeral. Stamens indefinite, with filiform filaments; anthers short. Style one. Capsule 3-valved.

1. HELIANTHEMUM, Tourn.

Petals broad. Stamens numerous (about 20). Style short; stigma 3-lobed. Low branching herbs, or somewhat woody; flowers yellow, opening only once, in sunshine.

1. **H. scoparium**, Nutt. Much branched, hairy or smooth, about a foot high; leaves narrow, 4 to 12 lines long, alternate; flowers on slender pedicels, one or several terminating the branches; petals 4 lines long.

Order 8. VIOLACEÆ.

Herbs distinguished by the irregular one-spurred corolla of 5 petals, 5 stamens, adnate introse anthers conniving over the pistil, which has a club-shaped style with a one sided

VIOLACEÆ. (VIOLET FAMILY.)

stigma, a one celled ovary, forming a capsule, which splits at maturity into three parts. Represented only by the familiar genus

1. VIOLA, L.

Sepals unequal, auricled at the base. Petals unequal, lower one spurred. Anthers nearly sessile, often coherent, the connectives of the two lower bearing spurs which are inclosed by the spur of the petal.

* *Leaves undivided.*

+ *Flowers not yellow, or orange.*

1. **V. canina**, L., var. **adunca**, Gr. Flowers violet or purple. Low stems sending out runners; leaves ovate, often somewhat cordate at the base, obscurely crenate; stipules foliaceous, narrowly lanceolate, lacerately toothed; spur as long as the sepals, curved; lateral petals bearded.

Var. **longipes**, Wat. The obtuse spur straight.

2. **V. ocellata**, Torr. & Gr. Stems nearly erect, 6 to 12 inches high; leaves cordate to cordate-ovate, acutish, conspicuously crenate; stipules small, scarious; upper petals white within, purple-brown without, the others pale-yellow veined with purple.

+ + *Flowers yellow, tinged with purple.*

3. **V. pedunculata**, Torr. & Gr. Stems with a decumbent or procumbent base; leaves rombic-cordate, with truncate or abruptly cuneate base, obtuse, coarsely crenate; stipules foliaceous, narrowly lanceolate, entire or gashed; showy flowers on peduncles exceeding the leaves; petals 6 to 9 lines long, the upper tinged with brown on the outside, the others veined with deep purple; lateral petals bearded; capsule smooth.

4. **V. aurea**, Kellogg. Leaves ovate to lanceolate, cuneate or sometimes truncate at base, obtuse, coarsely crenate; stipules foliaceous, lanceolate, laciniate; peduncle but little longer than the leaves; petals 4 to 6 lines long, as in the last, but lighter yellow; capsule pubescent.

5. **V. Nuttallii**, Pursh. Leaves oblong-ovate to oblong, attenuate into a long petiole, entire, or obscurely sinuate; stipules entire; peduncles usually shorter than the leaves.

+ + + *Flowers yellow.*

6. **V. sarmentosa**, Dougl. Leaves rounded-cordate, reniform, or sometimes ovate, finely crenate, usually punctate with dark dots. Flowers small.

* * *Leaves divided or lobed; flowers yellow, tinged with brown-purple.*

7. **V. lobata**, Benth. Distinguished by its stout stems and large palmately 5 to 9-lobed leaves. Flowers large.

8. **V. chrysantha**, Hook. Stems short; leaves bipinnatifid, with narrow segments. Flowers large, like **V. pedunculata**, but the lateral petals are not bearded.

CARYOPHYLLACEÆ. (PINK FAMILY.) 27

Order 9. **POLYGALACEÆ.**

Herbs or shrubs, with simple entire exstipulate leaves, remarkable for the papilionaceous-looking flowers. In our genus the ovary is 2-celled.

1. POLYGALA, Tourn.

Sepals 5, very unequal, the 2 lateral ones large and petal-like. Petals 3, united to each other and to the stamen-tube, the middle one hooded and often crested or beaked. Stamens 6 to 8, the filaments united below into a split sheath, adnate at the base to the petals. The 2-celled ovary forms a capsule flattened contrary to the partition, notched or retuse above.

1. **P. cucullata**, Benth. Stems slender from a woody base, 2 to 8 inches high; leaves smooth, oblong-lanceolate or ovate-elliptical, ½ to 1 inch long, short petioled; flowers rose-color; outer sepals 2½ lines long, rounded-saccate at the base; the wings broadly spatulate, 4 to 6 lines long.

2. **P. Californica**, Nutt. Stouter; flowers greenish white.

Order 10. **CARYOPHYLLACEÆ.**

Herbs with regular and mostly perfect flowers, persistent calyx, its parts and the petals 4 or 5 and imbricated or the latter sometimes convolute in the bud, the distinct stamens commonly twice as many as the petals, ovary 1-celled with a free central placenta. Stems usually swollen at the nodes. Leaves opposite, often united at the base by a transverse line, in one group with interposed scarious stipules. Styles 2 to 5, mostly distinct. Fruit a capsule opening by valves, or by teeth at the summit. Flowers terminal, or in the forks, or in cymes.

Many species in this order are difficult to determine.

* *Sepals united into a 4-5-toothed calyx. Petals long-clawed.*
Petals with bifid appendages ....Silene. 1

** *Sepals distinct; petals without claws.*
Petals bifid; capsule cylindric...Cerastium. 2
Petals bifid capsule globose...Stellaria. 3
Petals entire; capsule globose..Arenaria. 4
Stipules present; styles 5..Spergula. 5
Stipules present; styles 3...Lepigonum. 6

1. SILENE, L.

Calyx tubular, cylindrical to campanulate, 5-toothed, 10-nerved. Petals 5, with narrow claws; the blade mostly bifid or many-cleft and usually crowned with 2 scales at the base. Stamens 10; styles 3, erect. Capsule dehiscent by 6, rarely 3 teeth.

1. **S. Gallica**, L. Hairy; leaves spatulate, 1 to 1½ inches long; calyx oblong-cylindric, becoming expanded by the growth of the ovoid capsule; flowers small, rose-colored, in one-sided close racemes; petals entire, slightly twisted.

2. **S. Californica**, Durand. Glandular-pubescent; stems 6 inches to 3 ft. high, lax, leafy; flowers large, deep scarlet, few at the ends of the branches; calyx 7 to 10 lines long; petals deeply parted with bifid segments, the lobes 2-3-toothed or entire, with often a lateral one.

3. **S. Douglasii**, Hook. Stems simple few-flowered; leaves narrowly oblanceolate to linear, an inch or two long; calyx oblong-cylindric, often inflated, 5 to 7 lines long; petals rose-color or nearly white; 8 to 10 lines long, bifid with broad obtuse lobes; claw broadly auricled; capsule oblong-ovate, long stiped.

2. CERASTIUM, L.   Mouse-ear Chickweed.

Sepals 5. Petals 5, emarginate or bifid. Stamens 10. Styles 5, rarely less. The curved capsule dehiscing by twice as many teeth as there are styles. Flowers white.

1. **C. pilosum**, Ledeb. Erect, rather stout, more or less densely pilose; leaves oblong-lanceolate, ½ to an inch or more long, acute, almost sheathing at the base; flowers from ½ to 1 inch in diameter.

C. ARVENSE, L., has downy acute leaves.
C. VULGATUM, L.; has ovate or obovate obtuse leaves; flowers clustered.

3. STELLARIA, L.   Chickweed.

Sepals 5, rarely 4. Petals as many, 2-cleft. Stamens 10, or fewer by abortion. Low herbs with minute white flowers and 4-angled stems.

1. **S. media**, L. Weak and spreading, rooting at the lower joints; the ovate leaves less than an inch long on hairy petioles, or the upper ones sessile; stamens 3 to 10. Introduced from Europe.

2. **S. nitens**, Nutt., has small sessile lanceolate leaves and narrow shining sepals surpassing the minute petals.

3. **S. littoralis**, Torr., is rather a stout hairy plant, with ovate leaves; flowers in a terminal cyme. May be found on the sea-shore.

4. ARENARIA, L.   Sandwort.

Distinguished chiefly from *Stellaria* by the entire petals and usually by the tufted stems and subulate rigid leaves. In our species the 3 valves of the capsule are entire; bracts foliaceous.

1. **A. Douglasii**, Torr. & Gr. Slender, much branched, 3 to 6 inches high; leaves

filiform, 3 to 12 lines long; flowers on long slender pedicels; sepals 3-nerved; petals obovate, 2 lines long or more; longer than the sepals.

2. **A. Californica**, Brew. Leaves lanceolate, 1 or 2 lines long; flowers smaller than the last; petals spatulate.

3. **A. palustris**, Wat. Stems weak, 4 to 8 inches high; leaves linear, flaccid, 6 to 12 lines long; flowers few on long pedicels; petals 3 or 4 lines long. In swamps.

5. SPERGULA, L.   CORN-SPURRY.

Sepals 5. Petals 5, entire. Stamens 10, rarely 5. Ovary 1-celled, many-ovuled; styles 5, alternate with the sepals. Annuals dichotomously branched, with awl-shaped apparently whorled leaves (fascicled).

1. **S. arvensis**, L. The almost filiform leaves 1 or 2 inches long; flowers white, the long pedicels at length reflexed. Naturalized.

6. LEPIGONUM, Fries.   SAND-SPURRY.

Sepals 5. Petals 5, entire, rarely fewer. Stamens 10, or fewer by abortion. Ovary 1-celled, many ovuled; styles 3, or rarely 5. Low herbs, with setaceous or linear fascicled leaves; flowers white or pink, pedicelled.

1. **L. macrothecum**, Fisch. & Mey. Rather stout, often a foot high; leaves fleshy ½ to 2 inches long, with large ovate stipules; pedicels becoming reflexed; sepals 3 or more lines long, equaling the pinkish petals. In salt-marshes.

2. **L. medium**, Fries. More slender than the last, with smaller flowers on shorter pedicels.

ORDER 11. PORTULACACEÆ.

Succulent herbs, with simple and entire leaves, and regular but unsymmetrical perfect flowers; the sepals only 2, the petals 2 to 5 or more; the stamens opposite the petals when of the same number; the ovary 1-celled. Stamens sometimes indefinitely numerous, commonly adhering to the base of the petals, these sometimes united at the base. Style 2 to 8-cleft. Stipules none.

* *Sepals 2, distinct, persistent.*

Stamens more than 5 .... Calandrinia.   1
Stamens 5 .... Claytonia.   2
* * *Sepals 4 to 8* .... Lewisia.   3

1. CALANDRINIA, H B K.

Petals mostly 5 (3 to 10). Stamens 5 to 15. Ovary free, many-ovuled; style 3-cleft, short. Capsule globose or ovoid, 3-valved. Seeds shining-black. Low succulent herbs with alternate leaves.

30   HYPERICACEÆ. (ST. JOHN'S-WORT FAMILY.)

1. **C. Menziesii**, Hook. Smooth, branching from the base, the stems ascending; leaves linear to oblanceolate, 1 to 3 inches long, the lower on slender petioles; sepals keeled, the calyx 4-angled in the bud; petals broadly obovate, red to purple, 2 to 6 lines long. One of the most abundant of open ground early flowers.

2. CLAYTONIA, L.

Petals 5, equal. Stamens 5. Style 3-cleft. Capsule and seeds as in *Calandrinia*. Radical leaves numerous; cauline perfoliate, or a pair.

1. **C. perfoliata**, Donn. Stems 2 to 12 inches high; radical leaves long-petioled, broadly rhomboidal, or deltoid, or deltoid-cordate, ½ to 3 inches broad, obtuse; the cauline pair usually united to form an almost orbicular perfoliate leaf, concave above; the lax raceme of small pinkish flowers nearly sessile in the leaf-cup.

Var. **parviflora**, Torr. Radical leaves linear, or linear-spatulate.

Var. **spathulata**, Torr. Radical leaves linear; the cauline pair distinct or partly united on one side, ovate to lanceolate. Low and slender.

Var. **exigua**, Torr. Low, radical leaves narrowly linear or filiform; the cauline distinct, linear.

2. **C. Sibirica**, L. Stems 6 to 15 inches high; radical leaves lanceolate to rombic-ovate or nearly orbicular, long-petioled; the cauline pair ovate or varying from lanceolate to spatulate-obovate, sessile, distinct; raceme loose; the rose-colored or white petals 2 to 4 lines long.

3. LEWISIA, Pursh.

Petals 8 to 16, large and showy, rose-colored. Stamens numerous (40 or more). Style 3 to 8-paxted nearly to the base. Low acaulescent fleshy perennials, with fusiform roots, and short 1-flowered scapes.

1. **L. rediva**, Pursh. Leaves densely clustered, linear-oblong, subterete, 1 or 2 inches long, smooth and glaucous; scape jointed in the middle, bearing on the joint 5 to 7 subulate verticillate bracts; petals sometimes white, 8 to 16 lines long.

ORDER 12. **HYPERICACEÆ.**

Herbs or shrubs, with opposite entire punctate leaves, no stipules and perfect flowers with 4 or 5 petals and numerous stamens, the fruit a septicidal many-seeded capsule. Calyx of 4 or 5 persistent sepals. Filaments mostly in 3 sets. Styles 2 to 5, usually distinct.

1. HYPERICUM, L.   ST. JOHN'S-WORT.

Sepals and petals 5. The numerous stamens in three bundles. Ovary 1 to 3-celled, the ovules growing on the parietal placentæ. Flowers cymose, yellow.

1. **H. Scouleri.** Hook. Stems erect from a running rootstock ½ to 2 feet high, terete, simple or sparingly branched; leaves ovate to oblong, clasping, an inch or less long; petals punctate, 3 to 5 lines long; capsule 3-celled.

2. **H. concinnum,** Benth. Stems from a woody base, 3 to 6 inches high; leaves from oblong to linear, acute, an inch long or less, not clasping, usually folded.

3. **H. anagalloides,** Cham & Schlecht. Stems numerous, weak, rooting at the lower joints, 1 to 10 inches long; leaves broadly ovate or elliptical, 2 to 6 inches long, obtuse, clasping; sepals exceeding the petals; capsule 1-celled.

Order 13. MALVACEÆ.

Herbs or shrubs with alternate stipulate leaves; distinguished by the valvate calyx, convolute petals, their bases or short claws united with the base of a column of many united stamens, these with reniform anthers. Calyx 5-cleft or parted, persistent, with sometimes a calyx-like involucel of bracts. Petals 5, usually withering without falling off. Pistil usually either a ring of ovaries around a projecting receptacle or a 3-10-celled ovary; styles united at least at the base. Leaves usually palmately ribbed. Flowers axillary.

1. LAVATERA, L. Tree Mallow.

Involucel 3 to 6-cleft. Stamineal column divided into numerous filaments. Styles filiform. Fruit depressed; the several carpels separating from the prominent axis, 1-seeded.

1. **L. assurgentiflora,** Kellogg. A shrub 6 to 15 ft. high; flowers 1 to 4 in the axils on drooping pedicels; petals rose-purple, 1 to 1½ inches long, with a broad truncate limb and narrow claws having a pair of dense hairy tufts at the base. Commonly cultivated, but a native (?) of this State.

2. MALVA, L. Mallow.

Involucel 3-leaved. Petals obcordate, small. Herbaceous. Otherwise as *Lavatera*.

M. borealis, Wallman. Annual; leaves round-cordate, crenate, 5-7-lobed; peduncles short; petals pinkish-white, 2 or 3 lines long.

Distinguished from the biennial *M. rotundifolia* by its short peduncles, small flowers and rugose carpels.

3. SIDALCEA, Gr.

Involucel none. Stamineal column double; the filaments of the outer series usually united into 5 sets, opposite the petals. Flowers in a terminal raceme or spike. Herbs.

Perennial.

1. **S. malvæflora,** Gr. Perennial, 1 to 3 ft. high; leaves on elongated petioles,

orbicular to semi-circular in outline; the lower toothed or cleft, the upper more narrowly and deeply, 5 to 9-lobed or parted; the segments sparingly toothed, often linear and entire; flowers in naked elongated racemes; bractlets small, lanceolate; pedicels short, naked; calyx often tomentose; petals emarginate, 6 to 12 lines long, purple; carpels smooth.

2. **S. humilis**, Gr. Much resembling the last, but lower, and often decumbent at the base; leaves smaller; flowers fewer and more scattered; calyx larger, 3 to 6 lines long; carpels reticulated and pubescent.

* * *Annual.*

3. **S. diploscypha**, Gr. Pubescent with long spreading hairs, 1 to 2 ft. high; leaves deeply 5-9-cleft with lobed segments; bractlets conspicuous, 5 to 7-parted, hispid; flowers nearly sessile in close 3 to 5-flowered clusters; petals 6 to 12 lines long, broad and emarginate.

4. **S. malachroides**, Gr. Stout, hirsute, 3 to 6 ft. high, tufted; leaves large; flowers small, white or purplish, nearly sessile in close terminal heads on the short leafy branches; petals narrowly obcordate; sets of stamens indistinct.

Order 14. **LINACEÆ.**

A small order represented and characterized by the one genus

1. **LINUM**, L. Flax.

Parts of the flower 5, except sometimes in the pistil. Filaments united at the base with commonly alternating teeth. Styles 5, or sometimes only 2 or 3, distinct or united. Stigmas capitate or oblong; ovary globose. Seeds twice as many as the styles. Herbs with sessile entire leaves without stipules, and cymose or panicled flowers.

§ 1. *Styles 5. Flowers blue.*

1. **L. perenne**, L. Smooth, 1 to 2½ ft. high, branching above, leafy; leaves linear to linear-lanceolate, 3 to 18 lines long, acute; stipular glands none; flowers on slender pedicels, scattered, large.

§ 2. *Styles 3; petals appendaged at base, with a tooth on each side and a third adnate to the inner face of the claw.*

* *Flowers yellow; pedicels short.*

2. **L. Breweri**, Gr. Smooth, slender, 3 to 8 inches high or more, few flowered at the summit; leaves linear-setaceous, 6 to 8 lines long; stipular glands conspicuous; petals 3 or more lines long.

* * *Flowers rose-purple to white.*

3. **L. congestum**, Gr. Nearly smooth, excepting the calyx, about a foot high;

stipular glands very small; flowers in close terminal clusters; petals about 3 lines long; capsule globose.

4. **L. Californicum**, Gr. Smooth, glaucous, 6 to 18 inches high; stipular glands conspicuous; flowers in small cymes or the lower solitary; petals 4 lines long, capsule acute, shorter than the calyx.

5. **S. spergulinum**, Gr. Smooth, 6 to 15 inches high; leaves without stipular glands; pedicels 3 to 6 lines long, and mostly solitary; sepals slightly glandular, minute; capsule obtuse, exceeding the calyx slightly.

Order 15. GERANIACEÆ.

Flowers perfect on axillary peduncles, regular (in our species) and symmetrical, the parts in fives. Stamens mostly in two sets, those alternate with the petals sometimes sterile. Ovary deeply 5-lobed, with a prolonged axis, or 5-celled.

§ 1. *Carpels 5, one-seeded, separating at maturity from the long central axis; the styles forming long twisted tails.*

Fertile stamens 10; tails of the carpels not bearded.................... **Geranium. 1**
Fertile stamens 5; tails of the carpels bearded......................... **Erodium. 2**
§ 2. *Carpels 5, one-seeded, fleshy, distinct*............................**Limnanthes. 3**
§ 3. *Carpels combined into a 5-celled ovary*.................................**Oxalis. 4**

1. GERANIUM, L. CRANESBILL.

Stamens 10 with anthers, a gland behind the base of each of the shorter 5; filaments bearded at the base. Ovary 5-lobed; style 5-lobed at the top; the roundish-oblong carpels splitting away from the persistent beaked axis. Leaves palmately lobed and mostly opposite, scarious stipules; swollen-jointed stems.

1. **G. Carolinianum**, L. Diffusely branched, pubescent; leaves 1 to 2½ inches in diameter, palmately 5-7-parted, the divisions cleft into linear lobes; petals rose-colored equaling the awned sepals, 2 or 3 lines long; carpels hairy; tails half an inch long.

G. incisum, Nutt., with large purple flowers, grows in the Sierra Nevada, and in Humboldt County.

2. ERODIUM, L'Her.

Characters as in the last; but the filaments dilated, the 5 opposite to the petals sterile and scale-like; carpels attenuate to a sharp bearded base; the tails long bearded on the inner side. Leaves commonly pinnate and bipinnately parted or lobed; peduncles umbellately 2-several-flowered with a 4-bracted involucre at the base of the pedicels; flowers small.

1. **E. cicutarium**, L'Her. (FILARIA OR PIN-CLOVER.) Hairy, much branched,

decumbent; leaves pinnate the leaflets laciniately pinnatifid with narrow acute lobes, the opposite leaves unequal; the long peduncles in the axils of the smaller leaves bearing 4 to 8-flowered umbels; the slender pedicels at length reflexed, the fruit still erect; the bearded carpels with spirally twisted tails.

2. E. moschatum, L'Her. (Musky Filaria.) Similar to the last but of a lighter green and the leaflets unequally and doubly serrate, not pinnatifid. Gives out a musky odor when wilted.

3. E. macrophyllum, Hook. & Arn. Leaves reniform-cordate, 1 to 3 inches broad; sepals broad, 5 to 6 lines long.

3. LIMNANTHES, R. Br.

Glands 5, alternating with the petals. Stamens 10. Style 5-cleft at the apex. Annual low diffuse herbs, with pungent juice, growing in wet places; leaves pinnate, without stipules; flowers yellowish-white or rose-colored, solitary on axillary peduncles.

1. L. Douglasii, R. Br. Glabrous, yellowish green, weak and succulent stems; leaflets incisely lobed; peduncles at length 2 to 4 inches long; sepals lanceolate, 3 to 4 lines long, half the length of the oblong or obovate, emarginate or truncate petals.

Var alba, Hartweg. Villous sepals; shorter, white petals.

4. OXALIS, L.

The parts of the flower in fives. Stamens 10; the filaments dilated and united below. Capsule columnar or ovoid, beaked with the short style. Low herbs with sour watery juice; leaves alternate or radical, digitately trifoliolate, leaflets obcordate.

1. O. Oregana, Nutt. (Redwood Sorrel.) Acaulescent, rusty-villous; rootstock creeping; leaflets broadly obcordate, 1 to 1½ inches broad; petioles 2 to 8 inches long; scapes equaling or exceeding the leaves, mostly 1-flowered; petals 6 to 12 lines long, white or rose-colored, often veined with purple.

2. O. corniculata, L. (Yellow Sorrel.) Distinguished by its slender branching stems, and smaller yellow flowers.

Order 16. RUTACEÆ.

Pellucid or glandular-dotted aromatic leaves, along with definite hypogynous stamens and definite seeds characterize this order, although some of the orange-tribe have many stamens.

1. PTELEA, L. Hop-tree.

Flowers polygamous. Sepals, petals and stamens 4 or 5; ovary with a short, thick

stipe, 2-celled; style short; fruit a broadly winged orbicular samara, 2-seeded. Flowers small, greenish-white, in terminal cymes or compound corymbs.
 1. P. angustifolia, Benth. A shrub 5 to 25 ft. high, with chestnut colored punctate bark; leaves 3-foliolate.

Order 17. CELASTRACEÆ.

Suffiiciently characterized by the genus

1. EUONYMUS, Tourn.

Sepals and petals 4 or 5, widely spreading; Stamens as many very short on an angled disk; ovary immersed in the disk, 3-5-valved, colored, often warty. Fruit a red aril. Shrubs, with 4-angled branches, opposite petioled exstipulate serrate smooth leaves, and flowers in loose cymes on axillary peduncles.
 1. E. occidentalis, Nutt. 7 to 15 ft. high; leaves ovate to oblong-lanceolate, acuminate, serrulate, 2 to 4 inches long; peduncles 1-4-flowered; flowers dark reddish-brown, 4 to 6 lines in diameter, the parts in fives.

Order 18. RHAMNACEÆ.

Shrubs or small trees, with simple undivided leaves, small and often caducous stipules, and small regular flowers, the stamens borne on the calyx and alternate with its lobes; ovary 2 to 4-celled. Flowers often apetalous; a conspicuous disk adnate to the short tube of the calyx; petals often clawed; style or stigma 2-4-lobed; fruit berry-like or dry, containing 2 to 4 seed-like nutlets.
Calyx and disk free from the ovary; filaments short; fruit berry-like..... Rhamnus. 1
Calyx and disk adherent to the ovary; filaments long; fruit dry........ Ceanothus. 2

1. RHAMNUS, L.

Small greenish flowers; calyx 4-5-cleft, with erect or spreading lobes, the campanulate tube persistent; petals 4 or 5 or none, on the margin of the disk; claws short; stamens 4 or 5; leaves evergreen.

§ 1. *Flowers diœcious, apetalous, solitary or fascicled in the axils.*

 1. R. crocea, Nutt. Much branched, 3 to 15 ft. high; leaves coriaceous, oblong or obovate to obicular, 3 to 18 lines long, acutely denticulate, usually yellowish brown or copper-colored beneath; fruit red.

§ 2. *Flowers mostly perfect in pedunculate cymes.*

 2. R. Californica, Esch. Spreading 4 to 18 ft. high; leaves ovate-oblong to ellip-

RHAMNACEÆ. (BUCKTHORN FAMILY.)

tical, 1 to 4 inches long, denticulate or nearly entire; petals very small, broadly ovate, emarginate; fruit blackish-purple.

2. CEANOTHUS, L.

Calyx 5-cleft; the lobes acute; disk thick adhering to the tube and to the ovary; petals on long claws, hooded; stamens 5; filaments long-exserted; ovary 3-lobed; style short, 3-cleft. The small flowers are in showy thyrsoid or cymose clusters. Species difficult.

§ 1. *Leaves 3-nerved.*

1. **C. thyrsiflorus**, Esch. (CALIFORNIA LILAC.) Smooth, 6 to 15 ft. high; branches strongly angled; leaves rather thick, oblong to oblong-ovate, 1 to 1½ inches long, usually smooth and shining above, canescent beneath; flowers bright blue in dense compound racemes, terminating the long and somewhat leafy peduncles.

2. **C. integerrimus**, Hook & Arn. Slender, 2 or 3 ft. high; branches round, usually warty; leaves thin, bright green, ovate to ovate-oblong, 1 to 3 inches long; thyrse large, white-flowered.

3. **C. dentatus**, Torr & Gr. Low, not rigid; leaves small glandular-serrate, fascicled, the margin strongly undulate or revolute, somewhat resinous; flowers blue, in small roundish clusters.

4. **C. sorediatus**, Hook & Arn. Rigid; inflorescence pubescent; leaves silky on the nerves, ½ to 1½ inches long; flowers blue in shortly peduncled simple racemes ½ to 2 inches long.

5. **C. divaricatus**, Nutt. Grayish, usually spinose; leaves small, not tomentose beneath; flowers light blue or white, in nearly simple often elongated racemes, 1 to 4 inches long; fruit resinous.

6. **C. incanus**, Torr & Gr. Spinose leaves, hoary beneath with a very minute tomentum, cuneate to cordate at base; flowers in short racemes, white; fruit resinously warty. A straggling shrub along creeks.

§ 2. *Leaves pinnately veined.*

7. **C. papillosus**, Torr. & Gr. More or less hispidly villous or tomentose, 4 to 6 ft. high; leaves glandular-serrulate, and the upper surface glandular-papillose, narrowly oblong, 1 to 2 inches long on slender petioles; flowers blue, in close clusters or short racemes, terminating slender naked peduncles; fruit not resinous.

§ 3. *Leaves small, often opposite, very thick, with numerous straight lateral veins; stipules mostly large and warty; flowers in sessile or shortly peduncled axillary clusters; fruit larger, with 3 horn-like or warty prominences below the summit.*

8. **C. crassifolius**, Torr. Erect 4 to 12 ft. high, the young branches white with a villous tomentum; leaves somewhat spinosely-toothed or rarely entire and revolutely margined; flowers light blue or white, in dense clusters.

SAPINDACEÆ. (BUCKEYE FAMILY.)

9. **C. cuneatus**, Nutt. Similar to the last, but less tomentose; leaves cuneate-obovate or oblong, retuse above, on slender petioles; flowers in looser clusters.

10. **C. rigidus**, Nutt. Erect, 5 ft. high, the branchlets tomentose; leaves 2 to 5 lines long, cuneate-oblong or broadly obovate, few toothed above, very shortly petioled; flowers bright blue.

ORDER **VITACEÆ** has but one representative; the well-known California wild grape, **Vitis Californica**, Benth., which is common on the woody banks of streams.

ORDER 19. SAPINDACEÆ.

Trees or shrubs, mostly with compound or lobed leaves, with unsymmetrical or irregular flowers; the order best characterized under its suborders.

Under the order proper belongs

1. ÆSCULUS, L. BUCKEYE.

Leaves opposite, palmately 4–7-foliolate. Calyx tubular, unequally 5-toothed. Petals 4 or 5, unequal, with claws. Stamens 5 to 7, exserted and often unequal. Ovary 3-celled; style long. Fruit a large leathery 3-valved pod.

1. **Æ. Californica**, Nutt. Leaflets, usually 5, smooth, oblong-lanceolate, acute, obtuse at base, slenderly petiolulate, serrulate, 3 to 5 inches long; flowers in a close finely pubescent thyrse which is 6 to 12 inches long; calyx 2-lobed, the lobes scarcely toothed; petals white or pale rose, half an inch long or more; stamens 5 to 7; anthers orange colored. Fruit pear-shaped, 1½ to 2 inches long, containing, usually, one seed.

SUB-ORDER. ACERINEÆ.

Flowers polygamous or dioecious, regular, often apetalous. Ovary 2-lobed and 2-celled, each 1-seeded cell producing a wing. Leaves opposite without stipules.

2. ACER, Tourn. MAPLE.

Leaves palmately lobed. Calyx colored. Petals, usually 5. Stamens 3 to 12 inserted with the petals on a lobed disk. Styles 2. Fruit divaricately 2-winged.

1. **A. macrophyllum**, Pursh. (LARGE-LEAFED MAPLE.) A tree 2 or 3 feet in diameter; leaves 6 to 10 inches in diameter, deeply 3–5-cleft; flowers fragrant, yellow, in crowded pendulous racemes; fruit densely hairy; the smooth wings 1½ inches long.

2. **A. circinatum**, Pursh. (VINE-MAPLE.) A shrub or small tree; leaves 3 to 5 inches broad, 7–9-lobed, lobes sharply serrate; flowers in corymbs loosely 10–20-flowered, on slender 2-leaved branchlets; sepals red or purple, exceeding the greenish petals; fruit smooth.

3. **NEGUNDO**, Mœnch. Box-Elder.

Flowers diœcious. Calyx minute. Petals and disk none. Stamens 4 or 5. Ovary and fruit as in *Acer*. Trees; leaves pinnate; sterile flowers on clustered capillary pedicels, the fertile in drooping racemes.

1. **N. Californicum**, Torr. & Gr. Usually a small tree; leaves 3-foliolate, villous; leaflets ovate or oblong, acute, 3 or 4 inches long, the terminal largest and 3-5-lobed or coarsely serrate, the lateral ones coarsely serrate; fruit pubescent; wings slightly spreading.

ORDER **ANACARDIACEÆ** is represented by **Rhus, diversiloba**, Torr. & Gr. (POISON OAK); and **R. aromatica**, var. **trilobata**, Gr. The former has striped whitish nutlets; the latter, not poisonous, has red nutlets.

ORDER 20. **LEGUMINOSÆ**.

The single and simple free pistil becoming a legume in fruit, the alternate leaves with stipules and, in our genera, the papilionaceous corolla with 10 stamens, mark this order, one of the largest and most important in the vegetable kingdom.

Flowers irregular. Calyx 3-5-cleft or toothed, persistent. Corolla of 5 petals, the upper larger and always external, covering the lateral pair in the bud, and these covering the lower pair, which are more or less united, forming a keel which encloses the stamens and pistil. Filaments 10, rarely 5, commonly united around the pistil, either all united or nine and the upper one free. Ovary forming a pod with a single row of seeds attached to one side; style usually inflexed or curved. In *Cercis* the upper petal is small and enclosed by the wings. In *Amorpha* there is but one petal.

§ 1. *Stamens distinct.*

Leaves digitately 3-foliolate. Herbs; yellow flowers.................Thermopsis. 1
Shrub; purple flowers..Pickeringia. 2
Leaves unequally pinnate; shrubby; 1 petal.........................Amorpha. 9

§ 2. *Stamens all united into a sheath.*

Anthers of two forms; leaves digitate, more than 3 leaflets..............Lupinus. 3
Anthers all alike; leaves pinnately 3-foliolate......................Psoralea. 8

§ 3. *Stamens diadelphous (2 sets, 9 and 1).*

* *Leaves 3-foliolate; pods small.*

Flowers capitate. Corolla persistent..................................Trifolium. 4
Flowers in axillary racemes or spikes. Pod globular, wrinkled.........Melilotus. 5
Flowers in axillary spikes. Pod one-seeded...........................Psoralea. 8
Pod spirally coiled or reniform......................................Medicago. 6

LEGUMINOSÆ. (PEA FAMILY.) 39

* * *Leaves unequally pinnate; leaflets entire; no tendril.*

Flowers umbellate or solitary, axillary.................................**Hosackia.** 7
Flowers white or pinkish. Pod short, prickly.....................**Glycyrrhiza.** 10
Pods mostly inflated or nearly 2-celled...........................**Astragalus.** 11

* * * *Leaves terminated by a tendril or bristle or an imperfect leaflet.*

Style filiform, hairy around the apex.....................................**Vicia.** 12
Style flattened dorsally toward the apex, hairy on the inner side, usually twisted half round..**Lathyrus.** 13

1. THERMOPSIS, R. Br.

Calyx companulate, cleft to the middle. Standard roundish, shorter than the oblong wings, the sides reflexed; keel nearly straight, its petals somewhat united, equalling the wings. Perennial herbs with the aspect of Lupine; leaflets entire; stipules foliaceous; flowers large in long terminal racemes, with persistent bracts.

1. **T. Californica**, Wat. Woolly-tomentose; stipules lanceolate; leaflets obovate to oblanceolate, an inch or two long; bracts ovate; pod hairy.

2. PICKERINGIA, Nutt.

Calyx campanulate, turbinate at the base, repandly 4-toothed. Petals equal; standard orbicular, the sides reflexed; wings oblong; keel petals oblong, distinct, straight, obtuse. A low stout much branched spinose shrub; leaves evergreen, small, nearly sessile, 1-3-foliolate, without stipules; flowers large, solitary, axillary, nearly sessile.

1. **P. montana**, Nutt. Spreading, densely branched, 4 to 7 ft. high, silky-tomentose or smooth; leaflets 3 to 9 lines long; flowers from light cinnamon-red to purple, 7 to 9 lines long; stamens persistent.

3. LUPINUS, L. LUPINE.

Calyx deeply bilabiate, bibracteolate. Standard broad, the sides reflexed; wings united at the ends, enclosing the incurved beaked keel. Stipules adnate to the petioles; leaflets entire. Flowers in terminal racemes, verticillate or scattered, bracteate.

A large and difficult genus.

* *Annuals.*

Ovules 2; bracts persistent; flowers in whorls; leaves long petioled, approximate; stout.
Long-villous; flowers mostly purple............................**L. microcarpus.** 15
Smoother; flowers yellow to white..............................**L. densiflorus.** 16

Ovules several; bracts deciduous; flowers in whorls; petioles 1 to 3 times the length of the leaflets.

Puberulent; leaflets broad, smoother above; bracts short..............**L. affinis.** 8

LEGUMINOSÆ. (PEA FAMILY.)

Villous; leaflets narrow, both sides pubescent.
Bracts elongated; flowers rather large.................................L. nanus. 9
Bracts short; flowers small, narrow.............................L. micranthus. 10
 Ovules several; bracts somewhat persistent; flowers scattered; petioles 1 to 4 times the length of the leaflets.
Slender; leaflets smooth above; bracts long....................L. leptophyllus. 11
Slender; leaflets linear; bracts short............................L. sparsiflorus. 12
Stout; leaflets truncate; bracts short............................L. truncatus. 13
Stouter; leaflets broad; bracts short; very hispid..............L. hirsutissimus. 14

* * *Perennials; herbaceous, tall; flowers large; ovules 8 to 12.*

Stout; long petioles; leaflets 10 to 16, very large.................L. polyphyllus. 4
Stout; short petioles; leaflets 7 to 10, large.........................L. rivularis. 5
Slender, decumbent; short petioles; leaflets small...................L. littoralis. 6
Stoutish, erect; short petioles; keel narrow, falcate.................L. albicaulis. 7

* * * *Perennials; shrubby, leafy, silky-pubescent.*

Leaflets narrowly lanceolate; flowers yellow......................L. arboreus. 1
Densely silky-pubescent; flowers blue to white..................L. Chamissonis. 2
Pubescence short, tomentose; shrubby at the base.................L. Douglasii. 3

1. **L. arboreus**, Sims. Often 4 to 8 ft. high; sulphur-yellow, fragrant flowers, verticillate in a loose raceme; pods large, pubescent, 10–12-seeded.

2. **L. Chamissonis**. Esch. Less shrubby, 1 to 4 ft. high; leaflets 7 to 9, cuneate obovate, a half to an inch long, very silky on both sides; bracts lanceolate, shorter than the calyx; flowers sub-verticillate, blue, violet, rarely white. A variety about San Francisco with long bracts.

3. **L. Douglasii**, Agardh. Slightly woody at base; pubescence short, tomentose or silky; leaflets 7 to 9, oblanceolate to cuneate-oblong, 1 to 1½ inches long, pubescent on both sides; bracts linear-setaceous, exceeding the calyx; flowers, blue or purple; calyx with long setaceous bractlets.

4. **L. polyphyllus**, Lindl. Stout, erect, 2 to 5 ft. high, sparingly villous; stipules large, triangular to subulate; leaves distant, long petioled; leaflets 2 to 6 inches long; racemes a foot or two long; flowers mostly scattered on long pedicels, blue, purple or white; bracts oblanceolate, equaling or shorter than the calyx; keel naked.

5. **L. rivularis**, Dougl. Stout, erect, 2 to 6 ft. high, nearly smooth; stipules subulate or setaceous; leaflets 7 to 10, about equaling the petioles, ½ to 5 inches long; raceme often 1 to 2 ft. long; bracts setaceous, exceeding the calyx; flowers purple or rarely white; keel slightly ciliate.

6. **L. littoralis**, Dougl. Stems slender decumbent or ascending, 1 or 2 ft. long; leaflets a half to an inch long, at least half as long as the petioles; flowers blue or violet, with some yellow, in short racemes; keel ciliate; calyx large, with small bractlets.

7. **albicaulis**, Dougl. Distinguished by its flowers; which are light-blue to white, the standard strongly reflexed, the margins cohering near the apex, naked, acute; the narrow keel very strongly falcate.

8. **L. affinis**, Agardh. Stem a foot high; leaflets broadly wedge-obovate, emarginate or obtuse, an inch long or more; the petioles twice longer; petals 5 lines long; the keel usually naked; bracts short.

9. **L. nanus**, Dougl. Slender stem 6 inches to a foot high, villous, often branching from the base; leaflets linear to oblanceolate, half to an inch long, the petioles 1 to 3 times longer; bracts exceeding the calyx; petals very broad, 5 to 6 lines long, bluish-purple, or at first nearly white; the standard shorter and usually marked with purple lines.

10. **L. micranthus**, Dougl. Similar to the last, but the flowers smaller, in usually shorter more dense racemes; bracts shorter than the calyx; petals 2 to 3 lines long, narrow.

Var. **microphyllus**, Wat. The lower and more hirsute form, with leaflets but 3 to 6 lines long.

Var. **bicolor**, Wat. Flowers larger, more like *L. Nanus*.

Var. **trifidus**, Wat. Very hairy; lower lip of the calyx 3-parted.

11. **L. leptophyllus**, Benth. Rarely branched, 1 or 2 ft. high, villous; leaflets narrowly linear on slender petioles; smooth above; bracts setaceous, much exceeding the calyx; petals 5 or 6 lines long, bluish-lilac, with a deep crimson spot upon the standard.

12. **L. sparsiflorus**, Benth. Very slender, sparingly branched, 1 to 1½ ft. high, villous, with spreading hairs; upper leaves very small; leaflets 5 to 9, linear, ¼ to 1 inch long; petals violet, 5 lines long, the standard shorter; pod half an inch long.

13. **L. truncatus**, Nutt. Stout, branched, 1 to 2 ft. high; leaflets linear, narrowed from the truncate or somewhat 3-toothed apex to the base, smooth above, ¾ to 1½ inches long, nearly equaling the petiole; petals deep-purple, 4 or 5 lines long, the standard shorter; pod about an inch long.

Here belongs L. STIVERI, Kellogg. A beautiful species of the Sierra Nevada, with yellow standard and rose-colored wings.

14. **L. hirsutissimus**, Benth. A foot high or more, very hispid, with spreading straight and viscid stinging hairs; leaflets broadly cuneate-obovate, obtuse or retuse, rarely acute, mucronulate; flowers in loose racemes, reddish-purple, 5 to 7 lines long.

15. **L. microcarpus**, Sims. Villous, with long hairs, 6 to 18 inches high; leaves approximate on long petioles; leaflets usually 9, cuneate-oblong, obtuse or emarginate, smooth above, 1 to 2 inches long; calyx densely villous, large; petals purple to white, 6 or 7 lines long; the hairy 1-2-seeded pods 8 lines long.

16. **L. densiflorus**, Benth. Much resembling the last; calyx smooth or finely pubescent; petals yellow or ochroleucous, rarely white or pink.

L. luteolus, Kellogg, may be found, distinguished by its more slender habit, smaller and fewer leaflets, and bracts exceeding the calyx.

LEGUMINOSÆ. (PEA FAMILY.)

4. TRIFOLIUM, L. CLOVER.

Calyx 5-cleft with nearly equal teeth, persistent. Corolla withering, persistent; wings narrow, keel short obtuse. Stamens usually diadelphous. Style filiform. Pod small and usually inclosed in the calyx, membranaceous, indehiscent or dehiscent at the ventral suture, 1 to 6-seeded. Herbs with leaves palmately 3 or rarely 5–7-foliolate; stipules adnate to the petiole; flowers in capitate racemes, spikes or umbels, rarely few or solitary; peduncles axillary or only apparently terminal.

All our species annual.

§ 1. *Heads not involucrate; ovules 2.*

* *Heads apparently terminal; flowers sessile, not reflexed; calyx teeth plumose, filiform.*

1. **T. Macræi**, Hook. & Arn. Somewhat villous, erect, 6 to 12 inches high; stipules ovate to lanceolate; leaflets obovate to narrowly oblong, obtuse or retuse, serrulate, about half an inch long; flowers dark purple, 3 lines long, in dense ovate long peduncled heads; calyx very villous; the straight teeth as long as the petals, often tinged with purple; pod 1-seeded.

Var. **dichotomum**, Brew. A taller and stouter form, with larger flowers in heads nearly an inch long; corolla more conspicuous, tipped with white.

* * *Heads axillary, small; flowers on short pedicels, at length reflexed; calyx teeth subulate; mostly smooth.*

2. **T. ciliatum**, Nutt. Erect, often 1 to 2 ft. high; leaflets similar to the last; corolla white or purplish, little exserted, 3 lines long; calyx tube campanulate; the lanceolate teeth very acute, rigid, the scarious margin rigidly ciliate.

3. **T. gracilentum**, Torr. & Gr. Erect, slender, a foot high or less; stipules lanceolate; leaflets cuneate oblong to ovate or obcordate, retuse, about half an inch long, serrulate; flowers pale rose-color or purplish on pedicels a line long or less; calyx campanulate, the subulate teeth nearly equaling the corolla.

4. **bifidum**, Gr. Exactly like the last, but the leaflets narrow, the sides sparingly toothed or entire, and all deeply notched or cleft at the apex.

§ 2. *Heads subtended by an involucre; peduncles axillary; flowers sessile, not reflexed.*

* *Involucre not membranaceous, deeply lobed, and the lobes laciniately and sharply toothed; corolla not becoming inflated.*

5. **T. involucratum**, Willd. Smooth; stems ascending, often a span high or more; leaflets mostly oblanceolate and acute at each end, a half to an inch long; flowers half an inch long, in close heads, purple or rose-colored; the narrow calyx teeth all entire; ovules mostly 5 or 6.

Var. **heterodon**, Wat. Heads larger and leaflets broader; some of the calyx teeth setaciously cleft.

6. **tridentatum**, Lindl. Smooth or glandular-puberulent, slender and usually erect,

a half to two feet high; leaflets linear to narrowly lanceolate, sharply serrate; heads rather large, the flowers 6 to 8 lines long, purple, often tipped with white; calyx strongly nerved; the rigid teeth usually shorter than the tube, abruptly narrowed into the spinulose apex, often with a stout tooth on each side; ovules usually 2.

Var. **obtusiflorum**, Wat. Stouter and often glandular-puberulent, with broader leaflets and larger flowers; calyx teeth entire.

7. **T. pauciflorum**, Nutt. Smooth, very slender; stems ascending or decumbent; leaflets obovate to oblanceolate or sometimes linear, half an inch long or less, serrulate; heads few flowered; involucre small; flowers 3 or 4 lines long, not much exceeding the calyx; deep purple to light rose-colored; calyx teeth subulate, entire; pod 2-seeded.

* * *Involucre membranaceous, at least at the base, less deeply lobed; corolla not inflated.*

8. **T. microcephalum**, Pursh. Villous, with soft hairs, slender, erect or decumbent; stems often a foot or two long; leaflets oblanceolate to obovate, usually retuse, serrulate; heads small, dense; involucre about 9-lobed, the lobes acuminate 3-nerved, entire; calyx hairy, nearly equaling the white or light rose-colored corolla; ovules 2; pod 1-seeded.

9. **T. microdon**, Hook & Arn. Resembling the last; involucre broader, nearly inclosing the head; its lobes about 3-toothed; calyx smooth.

* * * *Standard becoming conspicuously inflated and inclosing the rest of the flower; involucre nearly obsolete in No. 12.*

10. **T. barbigerum**, Torr. Somewhat pubescent; stems rather stout, decumbent or ascending, a span high or less; stipules scarious; involucre as broad as the heads, shortly lobed; calyx-tube short, membranaceous; its teeth setaciously awned, plumose, the lower usually exceeding the purple corolla, sometimes 3-parted; pod 2-seeded.

Var. **Andrewsii**, Gr. A stout villous form, the heads sometimes an inch broad; calyx teeth very long.

11. **T. fucatum**, Lindl. Smooth; stems stout and succulent, a foot or two high; stipules large and scarious, usually very broad and entire; leaflets obovate, $\frac{1}{2}$ to $1\frac{1}{2}$ inches long; heads large; involucre broad, deeply cleft; flowers often an inch long, pale rose-colored or purplish; 2 6-seeded.

12. **T. depauperatum**, Desv. Smooth, low, slender; heads only 3-10-flowered; involucre scarcely more than a scarious ring.

13. **T. amplectans**, Torr & Gr. Like the last; the involucre larger. Probably only a variety.

5. MELILOTUS, Tourn. SWEET CLOVER.

Flowers as in **Trifolium**, except that the petals are free from the stamens and deciduous. Pod 2-seeded.

1. **M. parviflora**, Desf. Annual, smooth, erect, often 2 or 3 ft. high; leaflets

44   LEGUMINOSÆ. (PEA FAMILY.)

mostly cuneate, oblong, obtuse, denticulate, an inch long or less; flowers yellow, a line long, iu slender axillary pedunculate racemes; pedicels a line long.

6. MEDICAGO, L.

Characters nearly as the last; style subulate; pod compressed, falcate, incurved or spirally coiled.

1. **M. sativa,** L. (LUCERN, ALFALFA.) Stems erect, 1 to 4 ft. high; from a deep perennial root, smooth; leaflets cuneate-oblong or oblanceolate, toothed above; flowers 3 or 4 lines long, racemed; pods numerous, spirally twisted, veined, smooth.

2. **M. denticulata,** Willd. BUR-CLOVER. Annual, nearly smooth, prostrate or ascending; leaflets cuneate-obovate or obcordate, toothed above; flowers small, yellow, usually 3 to 8 in an axillary cluster; pods spiral, armed with a double row of hooked prickles.

3. **M. lupulina,** L. Pubescent, procumbent; flowers very small, yellow, in short spikes; pods smooth, reniform, 1-seeded.

7. HOSACKIA. Douglas.

Calyx teeth nearly equal, usually shorter than the tube. Petals free from the stamens, nearly equal; standard ovate or roundish, the claw often remote from the others; wings obovate or oblong; keel somewhat incurved. Style incurved. Pod linear, sessile, several-seeded, partitioned between the seeds.——Herbaceous or rarely suffrutescent; leaves pinnate, 2-many-foliolate; stipules minute and gland-like, rarely scarious or foliaceous; flowers yellow or reddish, in axillary sessile or pedunculate umbels.

The flowers usually change to reddish or reddish-brown in drying. Matured pods are necessary for the determination of species.

§ 1. *Pod shortly acute, linear and many-seeded, straight, smooth; seeds suborbicular; flowers and fruit not reflexed; peduncles long; keel broad above mostly obtuse.*

Stipules large, foliaceous; villous, viscid.........................**H. stipularis.** 1
   Stipules scarious; smooth.
Bract small or none; wings usually white............................**H. bicolor.** 2
Bract 1-3-foliolate, at the umbel; keel and wings purplish............**H. gracilis.** 3
   Stipules reduced to blackish glands.
Appressed-pubescent; tall, stout; pod long, smooth.................**H. grandiflora.** 4
Flowers very small, solitary..**H. parviflora.** 5

§ 2. *Pod shortly acute, 3-7-seeded, straight; flowers small, mostly solitary; keel acute; stipules gland-like; villous.*

Blade of the standard cordate; leaflets 3 to 5; nearly smooth.........**H. parviflora.** 5

LEGUMINOSÆ. (PEA FAMILY.) 45

Flowers peduncled; corolla scarcely exceeding the calyx; leaves nearly
   sessile, 1-3-foliolate...................................**H. Purshiana.** 6
Flowers nearly sessile, not bracteate; corolla larger; leaves petioled, 3-5-foliolate; low.
Calyx-teeth about equaling the tube, pod 5-seeded...............**H. subpinnata.** 7
Teeth much longer than the tube; pod 2-4-seeded..............**H. Trachycarpa.** 8

§ 3. *Pod long-attenuate upward, incurved, pubescent; stipules gland-like; leaflets 3 to 7;
seeds 1 or 2; peduncles short or none; flowers and fruit reflexed.*

Somewhat woody; nearly smooth; stems angled; leaflets mostly 3, oblong to linear.
Umbels sessile; teeth narrow, erect...**glabra.** 9
Peduncles short or nearly wanting; teeth usually recurved........**H. cytisoides.** 10
Peduncles shorter; teeth short and blunt...........................**H. juncea.** 11

Very silky-pubescent or tomentose; stems herbaceous: pod pubescent, short; umbels
on short peduncles.

Very pubescent throughout; flowers 3 or 4 lines long.............**H. tomentosa.** 12
Less pubescent; stem smooth; flowers smaller...................**H. Heermanni.** 13

1. **H. stipularis,** Benth. Rather tall, stout, two feet high or more, glandular; leaflets 15 to 21, obovate oblong, acute and mucronate, a half to an inch long; stipules large ovate; often fragrant.

2. **H. bicolor,** Dougl. Smooth, erect and stout; leaflets 5 to 9, obovate or oblong, a half to an inch long; stipules rather large; peduncles longer than the leaves, 3-7-flowered, naked or sometimes with a small 1-3-foliolate bract at the summit; flowers nearly sessile yellow, the wings often white; pod slender nearly 2 inches long.

3. **H. gracilis,** Benth. Much like the last; usually low and slender, the weak stems a span high or more; umbel with a petioled 1-3-foliolate bract; flowers yellow, keel and wings purplish.

4. **H. grandiflora,** Benth. Stout, 1 to 5 ft. high, more or less appressed silky-pubescent; leaflets 5 to 7 on an elongated rachis, 6 to 9 lines long; peduncles elongated; umbel 3-8-flowered, usually subtended by a single leaflet; flowers nearly sessile, 6 to 11 lines long, yellowish or greenish white, often tinged with purple, pod slender, smooth.

5. **H. parviflora,** Benth. Smooth or nearly so, stems slender, ascending, a span high or less; leaflets 3 to 5, obovate and very small to narrowly oblong and 6 to 8 lines long; bract 1-3-foliolate; flowers about 2 lines long, yellow.

H. Purshiana, Benth. Silky-villous, rarely smooth, often a foot high or more; leaflets varying from ovate to lanceolate, 3 to 9 lines long; peduncles usually exceeding the leaves; the solitary flowers 2 or 3 lines long.

7. **H. subpinnata,** Torr. & Gr. Villous or smooth, decumbent, a span high or less; leaflets half an inch long or less; flowers 3 or 4 lines long; pod linear oblong, about 5-seeded.

46   LEGUMINOSÆ. (PEA FAMILY.)

8. **H. brachycarpa**, Benth. Resembling the last; softly villous; pod villous, 2-4-seeded.

9. **H. glabra**, Torr. Very nearly smooth; stems woody at base, 2 to 8 ft. long, erect or decumbent; leaflets oblong to linear-oblong, 3 to 6 lines long; umbels numerous, sessile; flowers 3 or 4 lines long; seeds 2.

10. **H. cytisoides**, Benth. Resembling the last; peduncles equaling or exceeding the leaves, or sometimes very short, usually with a 1-3-foliolate bract at the top; calyx-teeth attenuate, mostly recurved.

11. **H. juncea**, Benth. Somewhat shrubby, erect; leaflets obovate to oblong, 2 to 4 lines long; umbels on very short peduncles or sessile; flowers about 3 lines long; calyx 2 lines long or less; teeth short and blunt.

12. **H. tomentosa**, Hook & Arn. Very pubescent, weak and flexuose, prostrate or ascending, a foot or more long; leaflets 5 to 7, cuneate-oblong to obovate, acute, 3 to 6 lines long; umbels on short bracteolate peduncles, or the uppermost sessile; flowers 3 or 4 lines long; calyx half as long or more, very villous.

13. **H. Heermannii**, Durand & Hilgard. Less pubescent, much branched and spreading; leaflets smaller; flowers smaller.

8. PSORALEA, L.

Calyx lobes nearly equal, or the lower one longer; the two upper often connate. Keel broad and obtuse above, united with the wings. Stamens diadelphous or monadelphous. Pod ovate, indehiscent, 1-seeded, thick, sessile. Perennial herbs punctate with dark glandular dots. Leaves pinnately 3-foliolate. Stipules free.

* *Stems prostrate, creeping; leaves orbicular.*

1. **P. orbicularis**, Lindl. Petioles 6 to 12 inches long; the leaflets 2 to 4 inches across, slightly cuneate at the base; peduncles equaling or exceeding the leaves, bearing a close villous spike of large flowers; the lower tooth of the calyx much the longest and about equaling the purplish corolla; stamens diadelphous.

* * *Stems erect.*

2. **P. strobilina**, Hook & Arn. Two or three feet high; petioles 3 or 4 inches long; leaflets rombic ovate, softly pubescent beneath, about 2 inches long; stipules large, membranaceous; flowers in short oblong spikes, smaller than the last; stamens monadelphous.

3. **P. macrostachya**, D C. Three to even twelve feet high; leaflets ovate-lanceolate, an inch or two long or more; peduncles much exceeding the leaves; spikes cylindrical, silky villous, the hairs often blackish; the lower tooth of the calyx but little the longest, scarcely equaling the purple petals; tenth stamen nearly free.

4. **P. physodes**, Dougl. A foot or two high; nearly smooth, slender; leaflets

LEGUMINOSÆ. (PEA FAMILY.) 47

ovate, mostly acute, about an inch long; the white or purplish flowers in short, close racemes; calyx at length inflated; stamens monadelphous.

9. AMORPHA, L.

Calyx obconical, nearly equally 5-toothed; wings and keel wanting; the standard erect, folded together. Stamens slightly united at the base, exserted. Pod 1-2-seeded. Shrubs, glandular-punctate; the unequally pinnate leaves with the leaflets stipellate; flowers purple or violet in dense clustered terminal spikes.
1. **A. Californica**, Nutt. Three to eight feet high, puberulent; leaflets 5 to 7 pairs, oblong-elliptical, obtuse, mucronulate, an inch long; spikes 1 to 6 inches long.

10. GLYCYRRHIZA, L. LIQUORICE.

Flowers nearly as in *Astragalus*. Erect perennial herbs, glandular viscid; leaves unequally pinnate; stipules deciduous; flowers in dense axillary pedunculate spikes; root large and sweet.
1. **G. lepidota**, Nutt., var. glutinosa, Wat. Two or three feet high; flowers yellowish white or pinkish; the short peduncles covered with stout viscid hairs. Rare; on water courses.

10. ASTRAGALUS, Tourn. RATTLE-WEED.

Calyx 5-toothed. Corolla and its slender clawed petals usually narrow; keel obtuse. Stamens diadelphous. Legume very various, commonly turgid or inflated, one or both sutures usually projecting inward, frequently so much as to divide the cell into two. Seeds few or many on slender stalks, generally small for the size of the pod. Herbs, or a few woody at the base; with unequally pinnate leaves, and small flowers, chiefly in simple spikes or racemes from the axils.

A vast genus of five or six hundred species; about fifty on the Pacific coast. The fruit is needed for the determination of the species.

* *Root annual; pod not inflated, 2-celled.*
Pod wrinkled, 2-lobed, 2-seeded.................................A. **didymocarpus**. 1
Pod not wrinkled, several-seeded.....................................A. **tener**. 2

* * *Root perennial; pod bladdery-inflated, 1-celled.*
Stipe a little exceeding the calyx; pod with pointed ends...........A. **oxyphysus**. 3
Stipe much exceeding the calyx; pod obtuse, one-sided..........A. **leucophyllus**. 4

Stipe, none; pod large and very bladdery, many seeded; leaflets mostly in many pairs; spike or raceme many flowered.
Stipules distinct; pod rather firm walled.......................A. **Crotalariæ**. 5
Stipules united; pod thin..A. **Menziesii**. 6
Stipules membranaceous; corolla yellowish.....................A. **Douglasii**. 7

48   LEGUMINOSÆ. (PEA FAMILY.)

1. **A. didymocarpus**, Hook. & Arn. Slender from 3 inches to a foot high; leaflets 9 to 15, narrowly oblong to linear and more or less cuneate, deeply notched at the apex; small flowers white and violet; pod not over two lines long, short oval and deeply 2-lobed lengthwise.

2. **A. tener.** Gr. A span or so in hight; leaflets similar to the last, not so deeply notched or entire; pod about half an inch long, 5–10-seeded; corolla 4 or 5 lines long, bright violet to pale and violet-tipped.

3. **A. oxyphysus**, Gr. Canescent with very soft silky pubescence; stem erect, 2 to 3 ft. high; leaflets oblong an inch or less in length; peduncles much exceeding the leaves; corolla greenish-white 8 lines long; bladdery pod acuminate and tapering into the recurved stipe which a little exceeds the calyx.

4. **A. leucophyllus**, Torr. & Gr. Less canescent than the last; flowers about half an inch long; corolla yellowish-white; the thin pod unequal-sided, an inch and a half long on a filiform pubescent stipe of almost equal length.

5. **A. Crotalariæ**, Gr., var. **virgatus**, Gr. Smooth or the young parts villous; stems 2 or 3 ft. high, stout; stipules scarious, triangular or subulate, distinct; peduncles elongated; racemes virgate and loose, 4 to 10 inches long; the white flowers soon deflexed.

6. **A. Menziesii**, Gr. Villous with whitish hairs or soon green and almost smooth; stems sometimes decumbent, 1 to 4 ft. high; the lower stipules united opposite the leaf; inflorescence similar to the last but more dense; pod larger (an inch and a half or more long) and more bladdery.

7. **A. Douglasii**, Gr. Cinereous-puberulent, almost smooth in age, stems ascending, a foot or so in height; leaflets in numerous pairs; linear or linear-oblong, 4 to 9 lines long; spike, half to an inch long; 10–20-flowered; pod gibbous-ovoid, 1½ to 2 inches long.

11. **VICIA**, Tourn.   VETCH. TARE.

Calyx 5-toothed or cleft, usually unequally. Wings adherent to the middle of the short keel. Stamens diadelphous or nearly so. Style filiform, inflexed, the apex surrounded by hairs or hairy upon the back. Pod flat 2-valved, shortly stiptate. Herbs, with angular stems climbing by branched tendrils terminating the pinnate leaves; leaflets entire or toothed at the apex; stipules semi-sagittate; flowers solitary or in loose axillary racemes.

* *Perennials; flowers in pedunculate racemes.*

1. **V. gigantea**, Hook. Stout and tall, climbing several feet high; leaflets 10 to 15 pairs, oblong, obtuse, mucronate, an inch or two long; stipules large; peduncles 5–18-flowered; corolla 6 or 7 lines long, pale purple; pod broadly oblong, 1½ inches long or more, smooth 3–4-seeded.

The seeds are large and edible; blackens in drying.

2. **V. Americana**, Muhl. Usually rather stout, 1 to 4 ft. high, smooth: leaflets 4 to 8 pairs, variable, linear to ovate-oblong, truncate to acute, ½ to 2 inches long; pedun-

cles 4–S-flowered; flowers purplish, 6 to 9 lines long; style very villous at the top; pods an inch long or more, 3–6-seeded.

Var. **truncata**, Brewer. Somewhat pubescent; leaflets truncate and often 3–5-toothed at the apex.

Var. **linearis**, Watson. Leaves all linear. Only the varieties are likely to be found.

* * *Slender annuals; flowers mostly solitary.*

3. **V. exigua**, Nutt. A span to two feet high, somewhat pubescent; leaflets about 4 pairs, linear, acute, a half to an inch long; peduncles usually short, rarely 2-flowered; flowers 3 lines long, purplish; pod about 6-seeded.

4. **V. sativa**, L. Rather stout, somewhat pubescent; leaflets 5 or 6 pairs, obovate-oblong to linear, retuse, long-mucronate; flowers nearly sessile, an inch long, violet-purple.—The common tare of Europe. Introduced.

12. LATHYRUS, L.

Style dorsally flattened toward the top, and usually twisted, hairy on the inner side. Peduncles usually equaling or exceeding the leaves and several flowered.

* *Rachis of the leaves tendril bearing; pod sessile; racemes several flowered.*

1. **L. venosus**, Muhl., var. **Californicus**, Watson. Very stout, several feet high; stems often strongly winged; leaflets oblong-ovate, acute; flowers nearly or quite an inch long, purple; pod about 2 inches long.

2. **L. vestitus**, Nutt. Slender, a foot to 6 or 10 feet high; stems not winged; stipules narrow, often small; flowers pale rose-color or violet, usually 7 to 10 lines long; ovary pubescent.

3. **L. palustris**, L. Slender, a foot or two high; stem often winged; leaflets narrowly oblong to linear, acute, an inch or two long; flowers purplish, half an inch long.

Var. **myrtifolius**, Gr. Stipules broader; leaflets ovate to oblong, shorter.

* *Rachis of the leaves not tendril bearing, or rarely so; pod shortly stipitate, peduncles long; 2–6-flowered.*

4. **L. littoralis**, Endl. Densely silky-villous throughout; stems numerous, from creeping root-stocks, stout, decumbent or ascending, $\frac{1}{2}$ to 2 ft. high; leaflets 1 to 3 pairs, with a small linear or oblong terminal one; calyx teeth nearly equal; standard bright purple, 6 to 8 lines long, exceeding the paler wings and keel; pod villous, an inch long.

Order 21. ROSACEÆ.

Herbs, shrubs or trees, with alternate leaves, usually evident stipules, mostly numerous stamens borne on the calyx; distinct free pistils from one to many, or in one sub-

50 ROSACEÆ. (ROSE FAMILY.)

order few and coherent with each other and adherent to the calyx forming a 2-several celled inferior ovary.
Nearly all the cultivated fruits of the temperate zones belong to this order.

SUB-ORDER 1. **AMYGDALEÆ**.

Carpels solitary, or rarely 5, becoming drupes, entirely free from the calyx, this or its lobes deciduous.——Trees or shrubs with bark and seeds tasting and smelling like those of the peach or cherry. Stipules few, deciduous.
Flowers perfect; carpel solitary..Prunus. 1
Flowers not all perfect; carpels 5..Nuttallia. 2

SUB-ORDER 2. **ROSACEÆ** PROPER.

Carpels free from the persistent calyx becoming akenes, follicles or berries.

§ 1. *Carpels few, becoming follicles; calyx open.*

Shrubs; follicles 2 to 8; flowers minute, in panicles.........................Spiræa. 3
Shrubs; follicles 1 to 5; flowers larger, in corymbs.........................Neiléia. 4

§ 2. *Carpels several or numerous, on a spongy receptacle, forming a compound berry*...Rubus. 5

§ 3. *Carpels one or many, becoming dry akenes.*

Shrubs; solitary, axillary apetalous flowers..........................Cercocarpus. 6
Herbs; carpels many, on a fleshy receptacle..............................Fragaria. 7
Herbs; carpels many, on a dry receptacle—
  Stamens 20 to 25...Potentilla. 8
  Stamens 10...Horkelia. 9
Shrub: heath-like, with subulate fascicled leaves...................Adenostoma. 10
§ 4. *Erect shrubs; showy flowers*..Rosa. 11

SUB-ORDER 3. **POMEÆ**.

Carpels 2 to 5, inclosed in and mostly adnate to the fleshy calyx-tube, in fruit becoming a berry-like pome. Trees or shrubs, with free stipules.
Stamens 10, in pairs; fruit red..Heteromeles. 12
Stamens 20; fruit black...Amelanchier. 13

1. **PRUNUS**, Tourn.   PLUM, CHERRY, ETC.

Calyx 5-cleft, deciduous. Petals 5, spreading. Stamens 15 to 25, inserted with the

ROSACEÆ. (ROSE FAMILY.) 51

petals. Ovary solitary, free, with two pendulous ovules; style terminal. Fruit a drupe, with usually a long stone containing one seed.
Deciduous; flowers white.
Corymbose; appearing before or with the leaves..................P. emarginata. 1
Racemose; appearing after the leaves.............................P. demissa. 2
Evergreen; leafless racemes axillary.................................P. ilicifolia. 3

1. **P. emarginata**, Walp. Four to eight feet high, with bark like the ordinary cherry tree, and chestnut-brown very slender branches; leaves oblong-obovate to oblanceolate, obtuse, narrowed to a short petiole; corymb 6-12-flowered, shorter than the leaves; flowers 4 to 6 lines broad; fruit globose, black; stone with a thick grooved ridge upon one side.

2. **P. demissa**, Walp. (WILD CHERRY.) Slender, 2 to 12 ft. high; leaves ovate to oblong-ovate, abruptly acuminate, mostly rounded or somewhat cordate at the base; racemes 3 or 4 inches long; fruit purplish-black or red, edible but astringent.

3. **P. ilicifolia**, Walp. (EVERGREEN CHERRY.) Much branched, 8 to 12 ft. high, with grayish-brown bark; leaves thick and rigid, shining above, broadly ovate to ovate-lanceolate, spinosely toothed; flowers small in racemes ½ to 2 inches long; fruit red or dark purple, half an inch or more thick.

2. NUTTALLIA, Torr. & Gr.   Oso Berry.

Petals 5, broadly spatulate, erect. Stamens 15 in two rows, 10 inserted with the petals, and 5 lower down upon the disk lining the calyx-tube, filaments very short, the lower declined. Carpels 5, inserted on the persistent base of the calyx-tube, free, smooth.

1. **N. cerasiformis**, Torr. & Gr. A shrub 2 to 15 ft. high; leaves rather broadly oblanceolate, short petioled; racemes of greenish white flowers, appearing with the branchlets from the same bud; drupes blue-black; with a slight furrow on the inner side, 6 to 8 lines long, bitter.

3. SPIRÆA, L.

Calyx persistent, 5-lobed. Petals 5, rounded, nearly sessile. Stamens 20 or more, inserted with the petals. Carpels distinct and sessile, becoming several-seeded follicles.

1. **S. discolor**, Pursh. A diffuse shrub, 4 ft. high or more with grayish brown bark, pubescent; leaves broadly ovate, truncate at base or cuneate into a slender petiole, pinnately toothed or lobed, the lobes often dentate; panicle of dingy white flowers much branched, tomentose.

Var. **ariæfolia**, Wat. Taller, 5 to 15 ft. high, leaves 2 or 3 inches long, panicle larger.
Var. **dumosa**, Wat. Only 1 or 2 ft. high, leaves an inch long or less, cuneate into a short margined petiole.

4. NEILLIA, Don.   Nine-Bark.

Carpels 1 to 5, inflated and divergent; flowers large, white, in simple corymbs.

1. **N. opulifolia,** Benth. & Hook. A shrub 3 to 10 ft. high, with slender spreading or recurved branches and ash-colored shreddy bark; leaves ovate to cordate, 3-lobed and toothed, 1 to 3 inches long.

5. RUBUS, L.

Calyx persistent 5-lobed. Petals 5, conspicuous. Stamens numerous. Carpels numerous, on a convex receptacle, becoming small globose 1-seeded drupes, forming a compound berry.—Fruit edible.

§ 1. *Fruit with a bloom, separating from the receptacle when ripe.*

Leaves simple, palmately lobed; stem soft, woody....................**Nutkanus.** 1
Leaves 3-foliolate, or on the flowering branches simple, rarely 5-foliolate; stems soft, woody, prickly—
    Flowers large, red..**spectabilis.** 2
    Flowers white...**leucodermis.** 3
Stems herbaceous, trailing unarmed...................................**pedatus.** 4

§ 2. *Fruit persistent, black and shining; stems prickly, flowers white*.........**ursinus.** 5

1. **R. Nutkanus,** Moc. (THIMBLE-BERRY.) Stems erect, 3 to 8 ft. high; older bark shreddy, no prickles; leaves 4 to 12 inches broad; flowers large white, rarely rose-colored, an inch or more across; fruit red, large.

2. **R. spectabilis,** Pursh. (SALMON-BERRY.) Stems 5 to 10 ft. high, similar to the last, but armed with a few prickles. Distinguished by its large red flowers and cylindrical-ovoid yellow or purplish berries.

Var. **Menziesii,** Wat. Densely tomentose and silky.

3. **R. leucodermis,** Dougl. (RASPBERRY.) May be known by its leaflets, white-tomentose beneath, prickly stem, white flowers, and its yellowish red white-bloomed fruit.

4. **R. pedatus,** Smith. Stems slender pubescent; leaflets cuneate-obovate, an inch or less in length; flowers white; the at length reflexed sepals exceeding the petals; berry of only 3 to 6 large red pulpy drupelets.

5. **R. ursinus,** Cham. & Schl. (BLACKBERRY.) Stems weak or trailing, 5 to 20 ft. long; fruit oblong.

6. CERCOCARPUS, HBK.

Calyx narrow, tubular, the campanulate 5-lobed limb deciduous. Petals none. Stamens in 2 or 3 rows on the limb of the calyx. Carpels solitary. Fruit a villous akene, included in the enlarged calyx-tube, tailed with the elongated exserted plumose twisted style.

Evergreen shrubs or trees. **C. ledifolius,** Nutt. is the MOUNTAIN MAHOGANY of the Sierra Nevada. The following is found in the Coast Range.

1. **C. parvifolius**, Nutt. A shrub 2 to 10 ft. high, or rarely a tree, branching from a thick base. Tails of the fruit often 4 inches long.

7. FRAGARIA. Tourn. STRAWBERRY.

Calyx persistent; limb 5-toothed, with 5 alternate bractlets. Petals white, spreading. Stamens in one row. Carpels numerous, smooth; styles lateral short. Receptacle much enlarged in fruit, conical, scarlet, bearing the small akenes on its surface.

1. **F. Chilensis**, Ehrh. Densely villous, with silky hairs; leaflets thick, smooth above; flowers often an inch broad; fruit ovate; akenes deeply pitted.
2. **F. Californica**, Cham. & Schl. Somewhat villous; leaves thin, veiny; fruit small; akenes not in pits.

8. POTENTILLA, L.

Calyx as in *Fragaria*. Petals yellow, rarely white. Stamens 20 to 50, marginal in 1 to 3 rows. Carpels numerous. Akenes small, on a dry receptacle.

1. **P. glandulosa**, Lindl. Perennial, erect, a foot or more high; leaves pinnate; leaflets 5 to 9, rounded, ovate, coarsely serrate; flowers cymose; calyx 4 to 6 lines long, usually villous, with coarse hairs; bractlets shorter than the lobes; petals not exceeding the calyx; stamens 25 in one row.
2. **P. Anserina**, L. (SILVER-WEED.) White tomentose and silky-villous leaves, all radical, often a foot long or more; leaflets 3 to 10 pairs, with smaller ones interposed, oblong, sharply serrate, tomentose, at least beneath; flowers yellow, solitary, on scape-like peduncles.

9. HORKELIA, Cham. & Schl.

Petals obovate to linear, often clawed, white or pink. Stamens 10, in two rows; filaments more or less dilated; those opposite to the sepals broadest. Flowers cymose.

* *Bractlets nearly as broad as the calyx-lobes.*

1. **H. Californica**, Cham. & Schl. Glandular-pubescent; stems a foot high or more; leaflets 5 to 10 pairs, 3 to 8 lines long; calyx about equaling the spatulate petals.

Var. **sericea**, Gr. Canescent throughout, with a dense, silky pubescence; leaflets larger.

* * *Bractlets much narrower than the calyx-lobes.*

2. **H. tenuiloba**, Gr. Canescently villous, a foot high; leaflets 8 to 12 pairs, deeply incised, 2 or 3 lines long.
3. **H. Bolanderi**, Gr. Densely hoary-pubescent, cespitose, the stems 3 or 4 inches high, the numerous leaflets minute, with rounded lobes.

10. ADENOSTOMA, Hook & Arn. CHAMISO.

Calyx persistent, 5-lobed; tube obconical, 10-ribbed; lobes membranaceous, broad.

ROSACEÆ. (ROSE FAMILY.)

Petals 5, orbicular, spreading. Stamens 10 to 15, usually 2 or 3 together between the petals. Fruit a membranaceous akene, included in the indurated calyx-tube.——Evergreen shrubs, somewhat resinous; flowers small, white, in terminal, racemose panicles.

1. **A. fasciculatum**, Hook & Arn, A diffusely branching shrub, 2 to 20 ft. high, with reddish virgate branches and grayish shreddy bark; leaves fascicled, linear subulate, 2 to 4 lines long, usually channeled on one side, smooth.

Alchemilla arvensis, Scop., belongs here. Its minute, greenish, apetalous flowers are fascicled in the axils of the small leaves and inclosed by the cleft stipules. A small under herb, growing on sandy hillsides.

Acæna trifida, R. & Pav. Is another apetalous herb, with silky, villous leaves and stem rising from a woody caudex; 3 to 15 inches high. The leaves are pinnate, the leaflets pinnately cleft into 3 to 7 segments. The greenish flowers with purple stamens are in a crowded terminal spike. Habitat similar to the last.

11. ROSA. Tourn. ROSE.

It is not necessary to here characterize this well-known genus.

1. **R. Californica**, Cham. & Schl. Erect, 2 to 8 ft. high, sparingly armed with usually recurved prickles, tomentose; leaflets 2 or 3 pairs; calyx lobes tomentose, often glandular leafy; petals 6 to 9 lines long; fruit globose.

2. **R. gymnocarpa**, Nutt. Slender, 1 to 4 ft. high, armed with straight slender prickles or unarmed, smooth; leaflets 2 to 4 pairs, glandular; flowers solitary, rarely 2 or 3, rarely an inch in diameter; calyx lobes at length deciduous; fruit small, ovate or pear-shaped.

12. HETEROMELES, J. Rœmer. PHOTINIA.

Calyx 5-parted. Petals 5, spreading. Stamens in pairs, opposite the calyx-teeth. Fruit red, berry-like.——An evergreen shrub or small tree, with coriaceous, simple, sharply serrate leaves. Flowers white in terminal panicles.

1. **H. arbutifolia**, Rœm. Leaves dark green above, lighter beneath, narrowly to oblong lanceolate, acute at each end, 2 to 4 inches long, on short petioles, slightly revolute margins; fruit 2 or 3 lines in diameter.

Pirus rivularis, Dougl., the *Oregon Crab-Apple*, may be found in Sonoma County.

13. AMELANCHIER, Med. SERVICE-BERRY.

Calyx-tube campanulate; the limb 5-parted, persistent. Petals 5, oblong, ascending. Stamens 20, short. Carpels 3 to 5 inferior, becoming membranaceous and partially 2-celled; styles united below or distinct. Fruit berry like, globose.——Shrubs or small trees; leaves simple, serrate; flowers white, racemose; fruit purplish, edible.

1. **A. alnifolia**, Nutt. A shrub 3 to 8 ft. high; leaves broadly ovate, sometimes cordate at the base, serrate only toward the summit, $\frac{1}{2}$ to $1\frac{1}{2}$ inches long.

SAXIFRAGACEÆ. (SAXIFRAGE FAMILY.) 55

ORDER **CALYCANTHACEÆ**, is represented by *Calycanthus occidentalis*, Hook. & Arn., an erect shrub 6 to 12 ft. high, with opposite entire lanceolate leaves, 3 to 6 inches long and large solitary livid or purplish red flowers; sepals and petals numerous, linear-spatulate. The common name of the Eastern species—*Sweet-Scented Shrub*—is scarcely applicable to our species.

ORDER 22. SAXIFRAGACEÆ.

Herbs, shrubs, or small trees, distinguished from *Rosaceæ* by albuminous seeds; usually by definite stamens, not more than twice the number of the calyx-lobes; commonly by the want of stipules; sometimes by the leaves being opposite; and in most by the partial or complete union of the 2 to 5 carpels into a compound ovary. Seeds usually indefinite or numerous. Petals and stamens on the calyx. Styles inclined to be distinct. Only the *Hydrangieæ* have many stamens.

Tribe 1. SAXIFRAGEÆ. Herbs, leaves mostly alternate and without distinct stipules. Styles or tips of the carpels distinct. Fruit capsular or follicular.

* *Ovary with 2 or rarely more cells, or of as many distinct carpels.*

Stamens 10, rarely more...Saxifraga. 1
Stamens 5...Boykinia. 2

** *Ovary 1-celled.*

Stamens 10, included..Tellima. 3
Stamens 10, exserted..Tiarella. 4
Stamens 5, and styles 2...Heuchera. 5

Tribe 2. HYDRANGIEÆ. Shrubs, leaves opposite, simple, no stipules. Fruit capsular.
A tall shrub. Large white flowers....................................Philadelphus. 6
Low, scarcely shrubby. Small flowers.................................Whipplea. 7

Tribe 3. GROSSULARIEÆ. Shrubs, leaves alternate with stipules adnate to the petiole or wanting. Fruit a berry.
Calyx-tube adnate to the ovary...Ribes. 8

1. SAXIFRAGA, L. SAXIFRAGE.

Calyx 5-lobed, free, or its tube coherent with the lower part of the ovary. Petals 5. Fruit of 2 follicles, or a 2-lobed capsule.—In our species stemless; flowers white.

1. **S. Virginiensis**, Michx. Leaves thickish, oblong-ovate to spatulate-obovate, coarsely toothed or almost entire, an inch or two long and the margined petiole often as long; scape viscid pubescent, 4 to 12 inches high, at length loosely many flowered in a paniculate cyme; flowers small white.

2. **S. integrifolia**, Hooker. Larger; leaves shorter petioled; flowers in a thyrsiform panicle; calyx lobes reflexed.

3. **S. Mertensiana**, Bong. Scape and leaves from a scaly granulate bulb; leaves rounded and cordate on long naked petioles; crenately or incisely lobed, the lobes often 3-toothed at the end; 2 to 4 inches across; calyx free.

2. BOYKINIA, Nutt.

Calyx 5-lobed, adherent to the ovary. Petals 5, entire, closed. Stamens alternating with the petals. Ovary and capsule 2-celled.—Perennial herbs, with creeping rootstocks, simple leafy stems; the leaves alternate, round-reniform, palmately lobed and incised or toothed, the teeth with callous-glandular tips, and the petioles mostly with stipule-like appendages at the base.

1. **B. occidentalis**, Torr. & Gr. Smoothish, or with some rusty hairs; a foot or two high; leaves thin-membranaceous, 3-7-lobed; petals white, 2 or 3 lines long.

3. TELLIMA, R. Br.

Calyx campanulate or turbinate, 5-lobed; the base coherent with the lower part of the ovary. Petals 5, inserted in the throat or sinuses of the calyx, laciniate-pinnatifid, 3-7-lobed, or entire. Stamens 10, short. Ovary short, 1-celled, with 2 or 3 parietal placentæ; styles 2 or 3, very short; stigmas capitate. Capsule conical, slightly 2-3-beaked.—Perennials, with round-cordate and toothed or palmately divided chiefly alternate leaves, few on simple stems, their petioles with stipule-like dilations at the base, and the flowers in a simple terminal raceme; petals white or pinkish.

Petals laciniate-pinnatifid ....T. **grandiflora.** 1
Petals entire, spatulate-obovate..................................T. **Cymbalaria.** 2
Petals entire; pedicels very short................................T. **Bolanderi.** 3
Petals obtusely 3-lobed..T. **heterophylla.** 4
Petals acutely 3-lobed ....T. **affinis.** 5

1. **T. grandiflora**, Dougl. A foot or more high, from short stout tufted rootstocks, hirsute or pubescent; leaves lobed, 2 to 4 inches in diameter; flowers dull-colored.

2. **T. Cymbalaria**, Gr. Stem or scape filiform, 4 to 12 inches high, bearing mostly a pair of opposite 3-lobed or parted leaves; radical leaves somewhat 3-5-lobed, half an inch across, flowers few and slender pedicelled, white.

3. **T. Bolanderi**, Gr. Stems a foot or two high, 1-4-leaved; radical and lower leaves lobed, the upper 3-5-parted; petals rarely with a small tooth on each side, white.

4. **T. heterophylla**, Hook. & Arn. Stems slender, a foot or less in height 1-3-leaved; leaves similar to the last, but smaller; flowers fewer and smaller, sometimes flesh-colored.

5. **T. affinis**, Gr. Rougher-pubescent; stem and leaves similar to the last; calyx densely rough glandular-pubescent; petals 4 or 5 lines long, white or flesh-colored.

SAXIFRAGACEÆ. (SAXIFRAGE FAMILY.) 57

4. TIARELLA, L.

Distinguished by the minute, slender petals, long exserted stamens, and the very unequal horns of the 2-carpeled ovary.

1. **T. unifoliata**, Hook. Somewhat hairy; flowering stems 4 to 15 inches high, 1-3-leaved; leaves thin, cordate, 3-5-lobed, crenate-toothed; flowers small, panicled.

5. HEUCHERA, L. ALUM-ROOT.

Calyx tube coherent with the lower half of the ovary. Petals small, entire, clawed. Ovary more or less 2-beaked; the beaks tapering into either filiform long, or subulate shorter styles.—Herbs with small, dull-colored paniculate flowers. Scarious stipules adnate or distinct. Leaves round-cordate, obtusely lobed, crenate-toothed.

1. **H. micrantha**, Dougl. Scape, or few leaved flowering stems, a foot or two high; leaves 2 to 4 inches in diameter; calyx acute at the base, lobes erect; styles slender.
2. **H. pilosissima**, Fisch. & Mey. Very villous-pubescent or hirsute, with viscid hairs; calyx rounded or obtuse at the base, the broad, short lobes incurving, densely hairy; styles short.

6. PHILADELPHUS, L. MOCK ORANGE.

Calyx adhering to the ovary nearly or quite to the summit, persistent. Petals 4 or 5, large, obovate or roundish. Stamens 20 to 40. Styles 3 to 5, united at the base or nearly to the top.—Shrubs with opposite leaves and showy white flowers.

1. **P. Gordianus**, Lindl. Six to twelve feet high; leaves ovate to oblong-ovate, mostly coarsely-serrate, 2 to 4 inches long; flowers in loose clusters, which are leafy at the base; petals frequently an inch long.

7. WHIPPLEA, Torr.

Calyx lobes thin, white or whitish. Petals ovate or oblong. Ovary 3 to 5-celled. Styles distinct, subulate.—Small, trailing or diffuse, ours half shrubby plants, with opposite, short petioled, 3-ribbed leaves, no stipules and small white cymose-clustered flowers; peduncles naked, terminal.

1. **W. modesta**, Torr. Leaves membranaceous, ovate or oval, obtusely few-toothed or entire, an inch or less long. Flower 2 lines long, clusters close-flowered, fragrant.

8. RIBES, L.

Calyx tube adnate to the globose ovary and extended beyond it, the limb commonly petaloid. Petals erect, mostly smaller than the calyx-lobes. Stamens alternate with the petals. Berry crowned by the withered remains of the flower.—Shrubs with alternate palmately lobed leaves.

58   CRASSULACEÆ. (STONE-CROP FAMILY.)

§ 1. *Thorny under the fascicles.* GOOSEBERRIES.

Berry prickly .... R. **Menziesii.** 1
Berry smooth .... R. **divaricatum.** 2
Berry dry; flowers large, bright-red .... R. **speciosum.** 3

§ 2. *Thornless and prickless.* CURRANTS.

Flowers rose-red to white .... R. **sanguineum.** 4
Flowers golden yellow .... R. **aureum.** 5

1. R. **Menziesii,** Pursh. Calyx about half an inch long, purplish red; its oblong lobes spreading or recurved, longer than the funnelform tube, hardly longer than the stamens which surpass the whitish petals; berry thickly covered with prickles.

2. R. **divaricatum,** Dougl. Flowers one-third of an inch long; calyx livid-purplish or greenish-white; its lobes about twice as long as the fan-shaped white petals, these only one-third as long as the stamens and villous 2-cleft style.

3. R. **speciosum,** Pursh. Very tall; flowers 2 to 5 on a bristly-glandular peduncle, drooping, fuchsia-like, almost an inch long and stamens as much longer.

4. R. **sanguineum,** Pursh. Racemes drooping, many flowered; calyx prolonged beyond the ovary into a campanulate tube 2 or 3 lines long, about equaling the lobes.—Runs into indefinite varieties.

5. R. **aureum,** Pursh. Flowers golden yellow, spicy-fragrant, in 5–10-flowered, leafy-bracted racemes.

ORDER 23.   CRASSULACEÆ.

Succulent or fleshy plants, with completely symmetrical as well as regular flowers.

Parts of the flower each 4 to 7; stamens twice as many. Petals distinct .... **Sedum.** 1
Petals somewhat united .... **Cotyledon.** 2

1. **SEDUM,** L.   STONE-CROP.

Sepals 4 or 5 united at the base. Carpels distinct or rarely connate at the base.

1. S. **spathulifolium,** Hook. Stems ascending from a branched rooting caudex, 4 to 6 inches high; leaves obovate or spatulate, flat, 6 to 10 lines long; flowers secund in a forked cyme, nearly sessile, 3 lines long; petals yellow, lanceolate acute.

2. **COTYLEDON,** L.

Petals united into a 5-lobed pitcher-shaped or cylindrical corolla. Stamens 10, inserted on the corolla-tube. Carpels usually distinct.

1. C. **farinosa,** Benth. & Hook. Acaulescent, more or less mealy-pulverulent; rosulate leaves lanceolate, acuminate, the larger ones 2 to 4 inches long; flowering branches a span high with scattered broadly ovate to lanceolate clasping leaves. Flowers yellow.

2. **C. cæspitosa,** Hawworth. Similar to the last; smooth glaucous-green; flowering branches 6 to 12 inches high, with broadly triangular-ovate clasping leaves. The most common species.

TILLÆA MINIMA, Miers., a small herb 1 to 3 inches high with clusters of minute white flowers in the axils of the opposite leaves is a common under-herb in moist places; as is also *T. angusti-folia*, Nutt., only an inch high with solitary flowers.

ORDER **LYTHRACEÆ** is represented by *Lythrum alatum*, Pursh., var. *linearifolium*, Gr. An herb a foot or two high with angled stems and small deep purple 6-petaled flowers solitary in the axils of the entire sessile leaves.

ORDER 24. ONAGRACEÆ.

Herbs (shrubby exotics), with the parts of the flowers in fours, the calyx tube adnate to the ovary, the petals borne on its throat, and the stamens as many or twice as many. Style always single.

Aquatic stems creeping...Jussiæa. 1
Flowers scarlet, fuchsia-like..Zauschneria. 2
Flowers small, purplish, leaves mostly opposite......................Epilobium. 3
Anthers attached near the center....................................Œnothera. 4
Flowers purple, calyx lobes reflexed..................................Godetia. 5
Petals clawed, calyx-tube short.......................................Clarkia. 6
Petals clawed, calyx-tube filiform.................................Eucharideum. 7
Flowers purple in leafy spikes....................................Boisduvalia. 8
Flowers minute, white, parts in twos..................................Circæa. 9

1. JUSSIÆA, L.

The 4 to 6 herbaceous lobes of the calyx persistent. Petals as many, obovate, spreading, yellow. Stamens twice as many. Capsule clavate.

1. **J. repens,** L., Var. **Californica,** Wat. Characterized sufficiently by its creeping stems and its solitary axillary flowers nearly an inch in diameter.

2. ZAUSCHNERIA, Presl.

Tube of the calyx much prolonged beyond the globose ovary, colored, the 4-lobed limb with 8 small deciduous scales, 4 erect and 4 deflexed. Stamens 8, exserted.

1. **Z. Californica,** Presl. The scarlet fuchsia-like flowers over an inch long cannot be mistaken.

3. EPILOBIUM, L. WILLOW-HERB.

The seeds tufted with silky hairs in linear 4-sided, 4-valved capsules best mark this difficult genus.

ONAGRACEÆ. (EVENING PRIMROSE FAMILY.)

4. ŒNOTHERA, L.

Calyx tube more or less prolonged beyond the ovary; segments reflexed. Petals 4; in our species yellow. Stamens 8, equal, or those opposite to the petals shorter. Style filiform; stigma 4-lobed or capitate.

* *Acaulescent. Calyx-tube filiform above the under-ground ovary.*

Leaves ovate to lanceolate..Œ. ovata. 1
Leaves linear...Œ. graciliflora. 2

* * *Caulescent. Calyx-tube obconic; capsule sessile, linear.*

Leaves thick; flowers small; capsule thick..................Œ. cheiranthifolia. 3
Flowers large; petals with a spot at the base.......................Œ. bistorta. 4
Flowers small; capsule contorted..............................Œ. micrantha. 5

Slender, leafy annuals; leaves linear; flowers small; capsule narrowly linear.

Flowers rarely reddening...Œ. dentata. 6
Flowers usually reddening..Œ. strigulosa. 7

1. **Œ. ovata**, Nutt. The radical leaves 4 to 6 inches long; calyx-tube scape-like, 1 to 4 inches long.

2. **Œ. graciliflora**, Hook & Arn. Canescently villous; calyx-tube equaling the leaves, 6 to 18 lines long; petals obcordate, 3 to 5 lines long, smaller than the last.

3. **Œ. cheiranthifolia**, Horn. Canescently pubescent; stems decumbent or ascending, 2 ft. long or more; leaves oblong or narrowly oblanceolate, sometimes broadly ovate or cordate, ½ to 2½ inches long, mostly entire, the lower petioled, the upper often clasping; ovary and calyx villous; flowers 2 to 5 lines in diameter; capsule 4 to 8 lines long. Near the sea on drifting sands.

4. **Œ. bistorta**, Nutt. Less common than the last; distinguished by its petals, 4 to 6 lines long, usually with a brown spot.

5. **Œ. micrantha**, Horn. A variable species distinguished from the last by its flowers, only 2 to 4 lines in diameter, with the petals sometimes 3-lobed; and by the contorted slender capsules, 8 to 18 lines long.

6. **Œ. dentata**, Cav. A span high or less; leaves linear, sessile, denticulate, 6 to 18 lines long; petals rounded, 2 to 4 lines long; capsule slender, attenuate, an inch long or more.

7. **Œ. strigulosa**, Torr. & Gr. Like the last; the capsule obtuse, scarcely attenuate. More common than the last.

Œnothera, biennis, L., the *Evening Primrose*, if found, may be known by its tall, erect stem and large flowers.

ONAGRACEÆ. (EVENING PRIMROSE FAMILY.) 61

5. GODETIA, Spach.

Distinguished from Œnothera by the anthers not versatile, and flowers not yellow.

* Flowers in a strict, mostly compact spike; capsule ovate to oblong; stems leafy.
Petals deep purple..G. purpurea. 1
Petals rose-colored with a spot..................................G. lepida. 2
Petals bluish-purple, 3 to 5 lines long........................G. albescens. 3

* * Flowers in usually a loose spike or raceme, mostly nodding in the bud; capsule linear; leaves distant.

+ Capsule sessile; stigma-lobes purplish.
Ovary and capsule short, villous, 2-costate..................G. quadrivulnera. 4
Capsule puberulent, not costate..................................G. tenella. 5
+ + Capsule pedicellate, not costate, stigma-lobes mostly yellow......G. amœna. 6
Small, hispid...G. hispidula. 7
Small, petals 2-lobed..G. biloba. 8

1. **G. purpurea**, Wat. Mostly very leafy, a foot or two high, puberulent, the ovary densely villous; leaves oblong to oblong-oblanceolate, an inch or two long, entire, sessile; flowers mostly in a leafy terminal cluster; petals 4 to 6 lines long; style shorter than the stamens; stigma-lobes very short, purple; capsule 6 to 9 lines long, not costate.

2. **G. lepida**, Lindl. Canescently puberulent, the stem usually white and shining. Easily distinguished by its flowers; the rose-colored petals with a dark spot near the top 9 to 12 lines long.

3. **G. albescens**, Lindl. Smaller leaves than the last, and much smaller almost blue flowers. Rare.

4. **G. quadrivulnera**, Spach. Puberulent, ovary and capsule more or less villous; stems usually slender, a foot or two high; leaves linear-lanceolate or linear, sessile or attenuate to a short petiole, entire or slightly denticulate, an inch or more long; petals deep-purple or purplish, 3 to 6 lines long; stigma-lobes short, purple.

5. **G. tenella**, Wat. Chiefly distinguished from the last by the capsule, which is 8 to 14 lines long, with nearly flat sides.

6. **G. amœna**, Lilja. Petals and purple anthers, frequently rather villous, varying from nearly white to rose-color, with more or less of purple, 8 to 15 lines long; capsule attenuate at each end.

7. **G. hispidula**, Wat. Is about a span high, often but 1-flowered; leaves narrowly linear; purple petals, 6 to 12 lines long.

8. **G. biloba**, Wat. Petals 2-lobed. Foot-hills of the Sierra Nevada.

6. CLARKIA, Pursh.

Petals 4, with claws, entire, purple. Stamens 8. Stigma with 4, at length spreading, sometimes unequal lobes. Capsule linear, 4-angled. Annuals, with erect brittle stems and alternate leaves on short petioles.

1. **C. elegans**, Dougl. Stems from 6 inches to 6 feet high; leaves broadly ovate to linear, repandly toothed; petals rhomboideal; stigma-lobes equal; capsule nearly sessile.

2. **C. rhomboidea**, Dougl. Is smaller; leaves petioled; claws of the petals toothed; capsule short, pedicelled.

7. EUCHARIDEUM, Fisch. & Mey.

Distinguished from *Clarkia* by the filiform calyx tube prolonged above the ovary, and stamens only 4.

1. **E. concinnum**, Fisch. & Mey. Closely resembles *Clarkia rhomboidea* in habit and foliage, calyx-tube an inch long; petals 3-lobed. Common.

8. BOISDUVALIA, Spach.

Petals 4, obovate-cuniform, sessile, 2-lobed, purple to white. Anthers not versatile.— Leaves alternate, simple, sessile; the small flowers in leafy spikes; our species villous.

1. **B. densiflora**, Wat. Canescent; 6 inches to 2 ft. high; leaves lanceolate to linear-lanceolate, mostly denticulate, 1 to 3 inches long; the floral leaves usually short and broad; flowers in usually a close terminal leafy spike or numerous short lateral spikelets; petals 3 to 6 lines long.

2. **B. Torreyi**, Wat. Rather slender, a span or two high; leaves 4 to 9 lines long; the floral leaves scarcely smaller; flowers very small.

9. **Circæa, Pacifica**, Asch. & Magn. In moist woods. Distinguished by its small indehiscent pear-shaped fruit covered with bristles and thin ovate opposite leaves.

Order 25. LOASACEÆ.

Herbaceous plants with either stinging or jointed and rough-barbed hairs; no stipules, calyx tube adnate to the 1-celled ovary. Stamens usually very numerous.

1. MENTZELIA, L.

Calyx cylindrical to ovoid; the persistent limb 5-toothed. Petals 5 or 10. Stamens numerous, inserted below the petals on the throat of the calyx; filaments free or in clusters opposite the petals, filiform or the outer petaloid. Style 3-cleft, the lobes often twisted.—The leaves are alternate, mostly coarsely-toothed or pinnatifid; flowers white to yellow or orange.

1. **M. albicaulis**, Dougl. Slender, 6 to 12 inches high or more; leaves linear-lanceolate, pinnatifid with numerous narrow lobes, the upper leaves broader and often lobed

CORNACEÆ. (DOGWOOD FAMILY.)

at the base only; flowers near the ends of the branches; petals 5, spatulate or obovate 2 to 3 lines long; capsule 6 to 9 lines long.

2. **M. gracilenta**, Torr. & Gr. Stems similar to the last; petals obovate, abruptly acuminate, an inch long; capsule 12 to 15 lines long.

3. **M. lævicaulis**, Torr. & Gr. Stout 2 or 3 ft. high; leaves lanceolate 2 to 8 inches long; flowers sessile on short branches, very large, light yellow; petals acute, 2 to 2½ inches long.

ORDER **CUCURBITACEÆ** is represented by **Megarrhiza Marah**, Wat. (BIG-ROOT). The cucumber-like vines, often 10 or even 30 ft. long; the sterile flowers white in racemes 4 to 12 inches long; the fruit ovate oblong, more or less covered with weak spines inclosing several nut-like seeds. **M. Californica**, Torr., has stiffer spines on smaller fruit; the fertile flowers without abortive stamens.

ORDER **FICOIDEÆ** is represented by **Mesembryanthemum, æquilaterale**, Haw., a very fleshy herb, with opposite three sided leaves 1 to 3 inches long and solitary red flowers; the petals numerous, linear. On the sea shore **Mollugo, verticellata**, L., will scarcely be noticed.

ORDER 26. UMBELLIFERÆ.

Herbs with small flowers in umbels, stamens and petals 5, borne on a 2-celled ovary which in fruit splits into a pair of dry usually flat indehiscent carpels. Since the generic distinctions depend upon characters of fruit and seed difficult of determination, the plants of this order are not here described.

ORDER **ARALIACEÆ** is represented by *Aralia Californica*, Wat. (SPIKENARD.) Grows in woods, along streams. Herbaceous stems, 8 to 10 ft. high; the white flowers in panicles a foot or two long and more.

ORDER 27. CORNACEÆ.

Trees or shrubs, rarely herbs, with simple entire mainly opposite leaves, no stipules, and flowers in cymes, capitate clusters or spikes; the petals and stamens 4, epigynous; calyx adnate to the 1-2-celled ovary, which becomes a drupe or berry.

1. CORNUS, L.

Flowers perfect. Calyx minutely 4-toothed. Petals 4, oblong or ovate. Stamens 4, with slender filaments. Style slender; stigma capitate or truncate. Fruit ovoid or oblong.

* *Flowers greenish, in a close head, surrounded by an involucre of 4 to 6 large, white, petal-like bracts.*

1. **C. Nuttallii**, Audubon. Usually a small tree; the involucre of yellowish or

64   CAPRIFOLIACEÆ. (HONEYSUCKLE FAMILY.)

white, often reddish bracts, 1½ to 3 inches long, abruptly acute. Fruit a large cluster of crimson berries.

2. **C. Canadensis,** L. Stem simple, herbaceous, 3 to 8 inches high; leaves in a whorl of 6 at the top, and a pair below; the 4 bracts 4 to 8 lines long.

* * *Flowers white or cream colored, cymose, not involucrate.*

3. **C. Californica,** C. A. Meyer. A shrub, 6 to 15 ft. high, with smooth, purplish branches; leaves ovate acute, obtuse at the base, 2 to 4 inches long, lighter colored beneath, with loose, silky hairs; flowers in small, dense, round-topped cymes.

4. **C. glabrata,** Benth. Bark gray; leaves oblong to narrowly ovate, acute at each end, alike green on both sides; flowers in open, flat cymes.

GARRYA ELLIPTICA, Dougl. and *G. Fremontii*, Torr., diœcious shrubs, belong here. The evergreen coriaceous leaves are opposite on the 4-angled branchlets, the short petioles connate; the apetalous flowers in axillary aments. Leaves of the former elliptical, undulate margins; the staminate aments long; leaves of the latter ovate to oblong, not undulate, lighter green.

DIVISION 2. GAMOPETALÆ.

Order 28. CAPRIFOLIACEÆ.

In our species shrubs with opposite leaves, no stipules, the calyx adherent to the 2-5-celled ovary, the stamens as many as the lobes of the rotate or tubular corolla.

Corolla rotate, regularly 5-lobed; white............................ **Sambucus.** 1
Corolla bell-shaped, regularly 4-5-lobed, pinkish............... **Symphoricarpus.** 2
Corolla tubular, irregular... **Lonicera.** 3

1. SAMBUCUS, Tourn. Elder.

Calyx teeth corolla lobes and stamens 5. Stigmas 3 to 5. Berries really drupes.—— Shrubs whose rank shoots are filled with a pith, half an inch in diameter. Leaves pinnately 5-11-foliolate. Flowers small, in large compound cymes.

1. **S. glauca,** Nutt. Cyme flat, 5-parted; fruit black, with a white bloom.
2. **S. racemosa,** L. Cyme ovate or pear-shaped; fruit bright red. Rare.

2. SYMPHORICARPUS, Dill. (Snowberry).

Calyx 5-toothed, occasionally 4-toothed, persistent. Corolla nearly or quite regular, from open campanulate to salver-form, 5-4-lobed. Stamens as many as the lobes of the corolla, inserted on its throat. Fruit globular, white.—Low shrubs, with oval or ob-

long leaves, mostly entire; and 2-bracteolate flowers in axillary and terminal clusters; rarely solitary.
1.   S. racemosus, Mich.   Erect, smooth; corolla very villous within.
2.   S. mollis, Nutt.   Low, diffuse or decumbent, softly pubescent; leaves small; corolla slightly villous.

2. LONICERA, L.   HONEYSUCKLE.

Corolla tubular, the tube commonly gibbous at the base and irregularly lobed. Stamens 5 inserted on the tube of the corolla.   Style filiform; stigma capitate.
1.   L. hispidula, Dougl.   Stems disposed to twine; leaves mostly oval, the lower short petioled, the upper pairs commonly connate; foliaceous stipule-like appendages between the leaves common; flowers sessile in a terminal head, pink or yellowish; berries red or orange.   Variable.
2.   L. involucrata, Banks.   An erect shrub, 4 to 10 ft. high; leaves ovate-oblong to broadly lanceolate, thin petioled; flowers a pair on axillary peduncles; below them a conspicuous involucre of 6 bracts, tinged with red or yellow; berries purple-black.

ORDER 29. RUBIACEÆ.

Known by having opposite entire leaves with intervening stipules, or whorled leaves without stipules, along with an inferior ovary and regular 4-5-merous flowers; the teeth of the calyx sometimes wanting. Stamens alternate with the lobes of the corolla and borne on its tube, distinct.

1. CEPHALANTHUS, L.   BUTTON-BUSH.

Flowers in a dense spherical head. Calyx inversely pyramidal, 4-5-toothed. Corolla with a long, slender tube and a small 4-cleft limb. Stamens 4, borne on the throat of the corolla, short. Style very long and slender.—Shrub with opposite leaves and stipules, or in whorls of 3 or 4. Peduncles axillary; flowers white.
1.   C. occidentalis, L.   Leaves ovate or lanceolate, 3 to 5 inches long; flower heads an inch in diameter.

2. GALIUM, L.   CLEAVERS.

Limb of the calyx obsolete. Corolla rotate, 4-parted, rarely 3-parted. Styles 2. Ovary 2-lobed. Fruit twin, biglobular. Herbs, sometimes woody at the base, with square stems, whorled leaves and minute flowers.

Leaves in fours, hispid, ovate .... G. Californicum.   1
Leaves in fours and pairs, smooth........................... G. Nuttallii.   2
Leaves mostly in whorls of eight........................... G. Aparine,   3

COMPOSITÆ. (ASTER FAMILY.)

Leaves in fives and sixes; fruit hairy. .... **G. triflorum.** 4
Leaves 4, 5 or 6 in a whorl; flowers white. .... **G. trifidum.** 5
Leaves in fours, 3-nerved, lanceolate. .... **G. boreale.** 9

1. **G. Californicum,** Hook and Arn. Low, branching; sterile flowers terminal, in threes, corolla yellowish; fertile ones solitary, recurved in fruit; fruit purple.
2. **G. Nuttallii,** Gr. Leaves 2 to 5 lines long, thickish, varying from ovate-oblong to linear-lanceolate, margins ciliate; flowers solitary.
3. **G. Aparine,** L. The margins midrib, and angles of the branches armed with spinulose bristles; peduncles 1-2-flowered; fruit large, white.(?)
4. **G. triflorum,** Michx. Bright green, nearly smooth; leaves oblong-lanceolate, acute at both ends, the margins and midrib often beset with hooked bristles; peduncles once or twice 3-forked; with hooked bristles.
5. **G. trifidum,** L. Nearly smooth, except the roughened angles of the slender stems; leaves 3 to 9 lines long; lobes of the white corolla often only three; fruit smooth.
6. **G. boreale,** L. Cymes many flowered, in a thyrsiform panicle.

ORDER 30. **VALERIANACEÆ.**

Herbs with opposite leaves, no stipules; the distinct stamens fewer than the lobes of the corolla, and borne on its tube; the inferior ovary with two empty cells, and one containing a solitary ovule, ripening into a kind of akene.

1. **PLECTRITIS,** (Lindl.) DC.

Limb of the calyx obsolete. Tube of the corolla very gibbous, spurred at the base; the short limb bilabiate; upper lip 2-cleft, lower 3-cleft. Fruit winged by the open sterile cells. Flowers white, small.
1. **P. congesta,** DC. Corolla about 3 lines long; its spur much shorter than the tube.
2. **P. macrocera,** Torr. & Gr. Corolla smaller; its thick spur about the length of the body.

ORDER 31. **COMPOSITÆ.**

Flowers, usually many in a dense head, sessile, on a common receptacle, surrounded by a calyx-like involucre; the calyx reduced to hairs or scales, or obsolete; the corolla tubular, equally lobed, ligulate or bilabiate, the 5 stamens united by their anthers into a tube inclosing the 2-parted style; the ovary inferior forming in fruit an akene which is usually crowned with the persistent calyx (pappus).

This the largest of all the orders, is represented in California by over 500 species, 140 of which grow within the limits of this Flora. Although the flower heads are frequently large, the separate flowers, with but few exceptions, are too small to be examined without the aid of a microscope skillfully used. The order is, therefore, far too difficult for the beginner.

ORDER LOBELIACEÆ. *Downingia elegans*, Torr., and *D. pulchella*, Torr., are two beautiful plants (the flowers resembling the cultivated *Lobelias*) sometimes cultivated under the name *Clintonia*, which properly belongs to an endogenous herb. The former has light blue flowers; the latter, deep azure-blue; both with white or yellowish centers. May be found in wet places.

ORDER 32. CAMPANULACEÆ.

Herbs with alternate leaves without stipules and regular flowers, having the calyx adnate to the ovary, distinct stamens (5, rarely 4) inserted with the corolla, alternate with its lobes.—Calyx persistent. Stamens with introse anthers, opening in the bud. Style single, its upper portion beset with hairs which collect the pollen, its summit 2-5-lobed or cleft.

* *Ovary and capsule long and narrow.*

Capsule opening at the top; calyx-lobes long..........................Githopsis. 1
Capsule opening by 2 or 3 holes on the sides.......................Specularia. 2

* * *Ovary and capsule short and broad, or globular.*

Capsule bursting indefinitely; calyx-lobes broad..................Heterocodon. 3
Capsule opening on the sides by 3 to 5 holes; calyx-lobes narrow.......Campanula. 4

1. GITHOPSIS, Nutt.

Flowers all alike. Calyx with a clavate 10-ribbed tube, and 5 long and narrow foliaceous lobes. Corolla tubular-campanulate, 5-lobed. Filaments short, dilated at the base. Ovary 3-celled; stigmas 3. Capsule strongly ribbed, crowned with the rigid calyx-lobes of its own length or longer, opening between them by a round hole.

1. G. **specularioides**, Nutt. An inch to a span high; leaves lanceolate-oblong or linear, sessile, coarsely toothed; flowers erect, deep blue, usually with a white center; the ovate lobes of the corolla about equaling the rigid calyx-lobes.

2. SPECULARIA. Heister.

Flowers in our species of two kinds: the lower and earlier usually with no corolla, Calyx-tube prismatic or elongated-obconical; the lobes 5, narrow. Corolla short and broad, rotate when fully expanded, 5-lobed. Stigmas 3 or 2. Capsule opening by round holes on the sides.

1. S. **biflora**, Gr. Stems slender; leaves sessile, ovate or oblong, crenately toothed, the upper reduced to lanceolate bracts; flowers 1, rarely 2, in each axil, nearly sessile; the

ERICACEÆ. (HEATH FAMILY.)

lower mostly apetalous, with 3 or 4 short calyx-lobes; the upper with 5 longer calyx-lobes, which are shorter than the blue or purple corolla. Capsule with openings near the top.
2. **S. perfoliata,** A. DC. Stouter, with clasping cordate leaves.

3 HETEROCODON, Nutt.

Flowers of two sorts. Stamens and styles as in *Campanula*. Capsule 3-angled. Otherwise sufficiently characterized in the synopsis.
1. **H. rariflorum,** Nutt. A delicate annual, with leafy filiform stems, diffusely branching; the thin leaves clasping by cordate bases, coarsely toothed. Corolla blue.

4. CAMPANULA. Tourn. BELLFLOWER.

Flowers all alike. Calyx-lobes narrow. Corolla campanulate or near it, 5-lobed. Stamens 5; filaments dilated at the base. Capsule 3–5-celled, opening on the sides or near the base by 3 to 5 small uplifting valves leaving round holes.
1. **C. prenanthoides,** Dur. A foot or two high; stems several-flowered; leaves ovate-oblong or lanceolate, sharply serrate, sessile, or the lower short-petioled; lobes of the blue corolla narrowly lanceolate, widely spreading; style long exserted; capsule 5-ribbed.

Order 33. ERICACEÆ.

Woody plants or perennial herbs, with symmetrical and mostly regular flowers; the stamens as many or twice as many as the petals or lobes of the corolla, and inserted with but rarely upon it; the anthers 2-celled, and the cells opening by a terminal pore; the ovary with as many cells as the divisions of the corolla or calyx; the seeds small. Corolla generally gamopetalous, sometimes of distinct petals, the insertion and that of the stamens hypogynous, or when the calyx is adnate epigynous around an annular disk. Style single. Leaves simple.

Sub-order 1. VACCINIEÆ.

Shrubs. Ovary wholly or partly inferior. Fruit a berry, crowned with the vestiges of the calyx-teeth.....................................**Vaccinium.** 1

Sub-order 2. ERICINEÆ.

Shrubs or trees. Calyx free. Corolla gamopetalous (in our own species). Stamens hypogynous. Anthers introse in the bud.

ERICACEÆ. (HEATH FAMILY.) 69

* *Fruit a berry, or berry-like drupe; corolla-tube inflated or urn-shaped, 5-toothed.
Evergreen.*
Tree; ovary 5-celled; berry many-seeded............................Arbutus. 2
Shrub; ovary 5-10-celled; drupe few-seeded..................Arctostaphylos. 3
Shrub; low; berry purple-black..................................Gaultheria. 4

* * *Fruit a naked capsule; corolla funnelform or campanulate, large, 5-lobed.*
Shrubs, with showy flowers.................................Rhododendron. 5

SUB-ORDER 3. **PYROLEÆ.**

Calyx free. Corolla of 5 (rarely 4) separate petals. Anthers extrose in the bud, the pores downward; introse (by bending downward on the end of the filament) in the open flower, the pores upward.

Stem woody, leaves whorled..Chimaphila. 6
Flowers on a scape...Pyrola. 7

SUB-ORDER 4. **MONOTROPEÆ.**

Root-parasitic, scaly-bracted herbs, wholly destitute of green foliage.
Flowers racemose, corolla globular-ovate ....Pterospora. 8

1. **VACCINIUM**, L. BLUEBERRY, BILBERRY, ETC.

Calyx 4-5-toothed on the summit of the ovary. Corolla various. Stamens 8 to 10; the anthers with the two cells separate, tapering upward into a tube opening at the top. Style long.
1. **V. ovatum**, Pursh. (CALIFORNIA HUCKLEBERRY). Shrub, erect, 3 to 5 ft high; evergreen; leaves thick, shining, ovate, acute, serrate; flowers with the parts in fives, stamens 10; corolla campanulate, pink; berries purple-black.

2. **ARBUTUS**, Tourn. MADRONO.

Calyx 5-lobed. Corolla ovate, 5-toothed; the teeth recurved. Stamens 10, included; anthers flattened, furnished with a pair of reflexed awns. Style rather long; berry with a rough surface.
1. **A. Menziesii**, Pursh. A handsome tree, with smooth bark turning brownish-red, which exfoliates except on the trunks of the larger trees; corolla white; berries deep orange.

3. **ARCTOSTAPHYLOS**, Adan. MANZANITA.

Flowers like those of *Arbutus* (but occasionally 4-merous and 8-androus), except that the 5 to 10 cells of the ovary contain each a single ovule, and the berry-like fruit has 5

70   ERICACEÆ. (HEATH FAMILY.)

to 10 bony seeds.—The white or rose-colored flowers in terminal racemes; the bark smooth, exfoliating.

* *Ovary and depressed-globose fruit more or less pubescent; branchlets often hispid.*

1. A. Andersonii, Gr. Erect, 6 or 10 ft. high; branchlets minutely tomentose, hispid with long, white, bristly hairs; leaves thin-coriaceous, green, lanceolate-oblong or ovate lanceolate, with a strongly sagittate-cordate base, sessile or nearly so, mostly spinulose-serrulate; fruit nearly or quite half an inch in diameter, with viscid bristles.

2. A. tomentosa, Dougl. Leaves thick and very rigid-coriaceous, varying from oblong-lanceolate to ovate and even cordate, entire or rarely serrulate, usually becoming vertical, smaller than the last; flowers in very short clustered racemes; fruit not viscid.

* * *Ovary glabrous; no hispid hairs on the branches and petioles.*

3. A. pumila, Nutt. Erect, dwarf, less than a foot high, tufted; leaves broadest near the apex, less than an inch long.

4. A. pungens, HBK. Leaves commonly becoming vertical by a twist of the distinct or pretty long petiole, very rigid, often glaucous or pale, entire or with a few teeth, varying from oblong-lanceolate to oval; flowers on smooth pedicels; filaments ciliate, bearded; fruit yellowish, turning dull red. Very variable.

A. GLAUCA, Lindl., if found, may be recognized by its large fruit, with the seeds consolidated into one woody stone, half an inch in diameter. *A. bicolor,* Gr., is smaller and has small apparently one-seeded berries.

4. GAULTHERIA, L. WINTERGREEN. SALAL.

Calyx 5-cleft, generally colored like the corolla. Corolla 5-toothed. Stamens 10 included, similar to those of *Arbutus*. Capsule 5-lobed, 5-celled, many-seeded, inclosed in the calyx, which enlarges and makes a juicy berry-like fruit.

1. G. Shallon, Pursh. Shrubby, stems ascending a foot or two in height; leaves ovate or slightly cordate, 2 to 4 inches long, finely serrate, shining; flowers white or rose-colored, in glandular-viscid racemes.

5. RHODODENDRON, L.

Calyx very small. Corolla often slightly irregular. Stamens 5 to 10; filaments filiform. Style long, commonly declined or incurved. Shrubs with alternate, entire leaves, usually crowded on the flowering branchlets; the showy flowers in terminal umbels or corymbs from ample scaly buds.

1. R. occidentale, Gr. (AZALEA.) A deciduous shrub, 2 to 6 ft. high; leaves obovate-oblong, bright green and shining above; corolla minutely viscid-pubescent outside, white; the upper lobe yellowish inside; the narrow funnel-form tube equaling the deeply 5-cleft slightly irregular limb; stamens and style much exserted, curved.—The showy fragrant flowers are sometimes nearly three inches long; rarely pinkish.

R. CALIFORNICUM, Hook., is a larger evergreen shrub, with large bell-shaped rose-purple flowers; a true *Rhododendron*, probably not found south of Mendocino County.

6. CHIMAPHILA, Pursh. Pipsissewa.

Corolla of rotately spreading, orbicular and concave petals. Stamens 10. Style very short, inversely conical, nearly immersed in the depressed ovary; stigma broad, its border somewhat 5-crenate.

C. umbellata, Nutt. A nearly herbaceous evergreen, 6 to 18 inches high; the usually whorled leaves oblanceolate, bright green; peduncle bearing 3 to 7 white or flesh-colored, waxy flowers.—Mt. St. Helena, *Miss E. Sweet.*

7. PYROLA, Tourn.

Corolla of 5 concave and converging petals. Stamens as in *Chimaphila*. Style generally long; stigma 5-lobed or 5-rayed.—Low and smooth perennial herbs, with broad and petioled leaves, close to the ground, and more or less scaly-bracted scape bearing a simple raceme of white, greenish or rose-colored, nodding flowers.

1. **P rotundifolia**, L. Leaves orbicular, varying to round-obovate or round-reniform, on slender, naked petioles; scape 6 to 14 inches high; probably our plants are of the Var. **bracteata**, Gr. A large form, with leaves 2 or 3 inches long; scape often over a foot high.

2. **P. picta**, Smith. Leaves thick, coriaceous; pale, sometimes purplish below; commonly blotched with white, ovate to obovate and lanceolate-oblong, on short petioles, 1 to 2 inches long; smaller than the last.

8. PTEROSPORA, Nutt.

Calyx deeply 5-parted, short, persistent. Corolla withering-persistent, globular-ovate, with contracted mouth; the 5 very short lobes, recurved. Stamens 10, included, short; stigma 5-lobed.

1. **P. andromedea**, Nutt. A stout, purplish-brown or chestnut-colored and clammy-pubescent herb, 1 to 3 ft. high; raceme long, many-flowered; corolla white, 3 lines long.

SARCODES SANGUINEA, Torr. The Snow Plant of the Sierra Nevada belongs here.

Order 34. PLUMBAGINACEÆ.

Chiefly maritime herbs, with regular flowers, the parts in fives; the stamens opposite the petals. Calyx tubular or funnel-form, 5-plaited, 5-toothed, persistent. Corolla in our genera with the long-clawed petals scarcely united. Stamens adnate to the base of the petals.

Flowers in a globose head on a simple scape...................Armeria. 1
Flowers on a branching scape...................Statice. 2

PRIMULACEÆ. (PRIMROSE FAMILY.)

1. ARMERIA, Willd. THRIFT.

Calyx scarious, funnel-form. Styles 5, filiform. Stemless perennials, with linear grass-like leaves in close tufts, the naked scape bearing a head of rose-colored flowers.
1. A. vulgaris, Willd. Scapes a foot or two high. On sandy hills along the coast.

2. STATICE, L. MARSH-ROSEMARY.

Flowers in small spikes or clusters, crowded at the extremities of a branching scape; their structure nearly as in *Armeria*. Leaves commonly with a broad blade, tapering into a petiole.
1. S. Limonium, L. Leaves obovate-oblong; spikelets 2-3-flowered. Salt marshes.

ORDER 35. PRIMULACEÆ.

Herbs, with perfect, regular flowers, well marked, by having the stamens as long as the lobes of the corolla, and opposite to them, inserted on its tube, a single entire style and stigma, a one-celled ovary, and capsular fruit. Calyx 4-8-cleft, commonly 5-cleft, hypogynous.—Leaves simple; stipules none. In *Glaux* the corolla is wanting; stamens on the calyx alternate with its lobes.

* *Flowers umbellate on a naked scape.*
Corolla deeply 4-5-parted, the lobes reflexed......................Dodecatheon. 1

** *Flowers axillary, on leafy stems.*
Corolla 5-9-parted, rotate..Trientalis. 2
Corolla 5-parted; prostrate stems......................................Anagallis. 3
Corolla wanting; calyx colored..Glaux. 4

1. DODECATHEON, L.

Calyx deeply 5-cleft, the divisions reflexed in the flower, afterwards erect over the ovate or oblong capsule. Corolla with a very short tube, a dilated, thickened throat and an abruptly reflexed 4-5-parted limb; its divisions long and narrow, entire. Stamens inserted in the throat of the corolla, erect, cohering around the slender exserted style.— Acaulescent perennial smooth herbs, with a tuft of radical leaves. Corolla purple, pink, or rarely white. Frequently the parts are in fours.
1. D. Meadia, L. Leaves varying from obovate to lanceolate, entire or toothed; scape 3 to 15 inches high; umbel, 2-20-flowered. A variable species. Ours is chiefly the
Var. brevifolium, with leaves round-obovate or spatulate, less than an inch to an inch and a half long.

2. TRIENTALIS, L. STAR-FLOWER.

Calyx and wheel-shaped corolla about 7-parted. Filaments slender, spreading.—Low and glabrous perennials, with simple stems, which bear a whorl of leaves at the summit, in their axils slender peduncles supporting star-shaped, white or pinkish flowers.

1. T. Europæa, L., Var. latifolia, Torr. Stems 4 to 8 inches high, springing from a little tuber.

3. ANAGALLIS, Tourn. PIMPERNEL.

Divisions of the rotate 5-parted corolla broad. Capsule globose.—Spreading, prostrate herbs, with opposite or whorled leaves and axillary flowers.

1. A. arvensis, L. Leaves ovate, sessile, shorter than the peduncles, sometimes in threes; flowers scarlet, purple, or nearly salmon-colored, rarely blue.

4. GLAUX, L. SEA MILKWORT.

Calyx campanulate, 5-cleft; the lobes ovate, petal-like. Filaments rather shorter than the calyx. Style filiform; stigma capitate.

1. G. maritima, L. Low, glabrous; branching stems 3 to 9 inches high, leafy to the top; leaves commonly opposite, fleshy, oblong, half an inch or less long, minutely dotted; flowers axillary, almost sessile, white or purplish.

ORDER **OLEACEÆ** is represented by *Fraxinus Oregana*, Nutt., the OREGON ASH.

ORDER **APOCYNACEÆ** is represented by *Apocynum cannabinum*, L. (INDIAN HEMP.) An herb with milky juice, tough bark, opposite entire exstipulate leaves, regular flowers, the sepals, petals and stamens five, the latter borne on the corolla alternate with its lobes and conniving around the stigma. The commonly sessile, oblong leaves often 3 or 4 inches long. The greenish-white small flowers in close cymes. *A. androsæmifolium*, L., has smaller ovate leaves, conspicuously petioled; flowers rose-colored.

ORDER 36. ASCLEPIADACEÆ.

Herbs with milky juice, no stipules, and regular flowers, with the parts in fives, except that there are two carpels with distinct ovaries and a common stigma to which the stamens are attached; the latter (in our genera) with hood-like appendages. Leaves entire, generally opposite, sometimes whorled. Flowers usually in simple umbels. Fruit a pair of follicles. Seeds almost always with a coma of silky down.

1. ASCLEPIAS, L. MILKWEED.

The calyx and corolla deeply 5-parted; the small divisions reflexed; filaments short, crowned behind each anther with a conspicuous hood from the cavity of which rises the subulate and usually falcate horn; anthers with thin scarious tips inflexed

74 GENTIANACEÆ. (GENTIAN FAMILY.)

over the truncate summit of the stigma, their wing-like edges meeting and projecting between the hoods; pollen in 10 wax-like masses. Follicles ovate or lanceolate. Seeds numerous, flat, downwardly imbricated all over the large, soon detached placenta; the upper end with a long tuft of down (coma).—Hoods in our species erect and not exceeding the stamens and stigma.

1. **A. fascicularis,** Decaisne. Smooth, slender, 1 to 5 ft. high; leaves in whorls of 3 to 5, or some in pairs, linear and linear-lanceolate; flowers white or whitish; horns longer than the hoods.

2. **A. vestita,** Hook & Arn. White-woolly; leaves opposite, ovate-lanceolate or oblong-lanceolate, almost sessile; umbels almost sessile; flowers about half an inch long, the hoods flesh-colored.

2. GOMPHOCARPUS, R. Br.

No horn to the hood of the stamens; otherwise as *Asclepias*.

§ 1. *Hoods saccate, pointless, lower than the anthers, opening down the back, as if 2-valved.*

1. **G. tomentosus,** Gr. White-tomentose, closely resembling *Asclepias vistitia;* stem acutely angled; leaves ovate or oblong (about 4 inches long); corolla greenish-white or purplish.

2. **G. purpurascens,** Gr. Canescently puberulent; stems 4 to 12 inches high; leaves ovate and somewhat cordate, an inch or two long; flowers small; the corolla red-purple; the hoods white.

§ 2. *Hoods erect, open down the front, somewhat surpassing the anthers.*

3. **G. cordifolius,** Benth. Green and smooth, 2 or 3 ft. high; leaves ovate or ovate-lanceolate, with cordate clasping base, opposite, rarely in threes, 2 to 5 inches long; flowers large; corolla dark red-purple; the hoods purplish.

Order 37. **GENTIANACEÆ.**

Glabrous herbs, with colorless, bitter juice, entire opposite and sessile leaves, no stipules, perfect and regular flowers, stamens as many as the lobes of the corolla and alternate with them, inserted on the tube, the anthers free from the stigma; ovary 1-celled; style one or none; the stigmas commonly two. Calyx persistent.

§ 1. *Corolla withering-persistent. Leaves opposite or whorled, entire, sessile.*

Corolla salver-form, red; calyx 5-parted..................................**Erythræa.** 1
Corolla short, salver-form, yellow; caylx 4-toothed....................**Microcala.** 2
Corolla funnel-form, blue...**Gentiana,** 3

§ 2. *Corolla deciduous. Leaves alternate, with sheathing petioles.*

Flowers borne on a naked scape...**Menyanthes,** 4

POLEMONIACEÆ. (GILIA FAMILY.) 75

1. ERYTHRÆA, Pers.

Stamens inserted on the throat of the corolla; filaments slender; anthers oblong or linear, twisting spirally after shedding the pollen. Style filiform; stigma wedge-shaped or fan-like. Capsule oblong, tapering upward.—Corolla occasionally only 4-parted.

1. **E. trichantha**, Grise. A span or less high, branched; lobes of the rose-red corolla lanceolate, fully half the length of the tube at the time of expansion, 3 or 4 lines long; calyx-lobes filiform, 3-angled.

2. **E. Muhlenbergii**, Grise. Two inches to a span high, simple or branched; leaves oblong, half an inch long; lobes of the corolla oval, very obtuse, becoming oblong, rose-red.

2. MICROCALA, Link.

Anthers round-cordate. Stigma peltate-dilated, at length separating or separable into 2 plates.

1. **M. quadrangularis**, Grise. An inch or two high, filiform, simple and 1-flowered, or branched at the base, with 1 to 3 pairs of minute oval or oblong leaves; peduncles naked, square; calyx short, square; corolla saffron-yellow.

3. GENTIANA, L. GENTIAN.

Calyx 4-5-toothed or cleft. Corolla 4-5-lobed, often with plaited and toothed folds in the sinuses. Stamens included; anthers sometimes cohering. Style none or very short; stigmas 2, thin and flat.

1. **G. affinis**, Grise., var. **ovata**, Gr. A span to a foot or two high; leaves ovate or oblong; flowers mostly 5 or more, in a leafy thyrsus; corolla blue, an inch or more in length; appendages mostly 2-cleft or 2–4-cuspidate, shorter than the round-ovate lobes.

4. MENYANTHES, Tourn. BUCKBEAN.

The campanulate corolla densely white-bearded on the upper surface, the lobes with the margins turned inward in the bud.

1. **M. trifoliata**, L. The alternate leaves long petioled, 3-foliolate; scape terminated by a short raceme of white or pinkish flowers; anthers dark-brown, sagittate.—In shallow water or on wet ground.

Order 38. POLEMONIACEÆ.

Chiefly herbs with simple or divided leaves, and no stipules; all the parts of the regular flower five, except the pistil, which has a 3-celled ovary and a 3-lobed style. Calyx imbricated in the bud, persistent. Corolla convolute in the bud. Stamens on the corolla alternate with its lobes distinct; anthers introrse.—In *Gilia* the cells of the ovary and the stigmas are occasionally reduced to two.

Stamens unequally inserted and included in the narrow tube of the salver-
form corolla...Collomia. 1
Stamens equally inserted on the throat or tube of the corolla; filaments
not declined..Gilia. 2
Filaments more or less declined; otherwise as *Gilia.* Leaves all pinnate
and alternate; corolla short...Polemonium. 3

1. COLLOMIA, Nutt.

The throat of the corolla commonly enlarged. Stamens more or less exserted, with slender filaments, mostly glandular-viscid; with alternate leaves, or the lower opposite, various.

* *Leaves simple and sessile, entire, the lower ones opposite.*

1. **C. gracilis**, Dougl. A span or two high, in age much branched; the flowers at length somewhat scattered; leaves lanceolate or linear, or the lowest oval or obovate, an inch or less long; corolla rose-purple, turning bluish, less than half an inch long, narrow.

* * *Leaves deeply cleft or compound, the lower petioled; stems loosely branched.*

2. **C. gilioides**, Benth. A span to 3 ft. high; lower leaves simply pinnately parted into linear lateral lobes, or the terminal lobe oblong and toothed, upper leaves 3-5-divided; corolla pink or purplish, its slender tube about half an inch long, twice or thrice the length of the calyx; capsule globular, 3-seeded.

3. **C. heterophylla**, Hook. A span or two high, diffuse; leaves mostly pinnately parted or the upper pinnatifid, and the lobes incised or cleft; the upper most often entire and broader, subtending the capitate-clustered flowers; corolla purplish, half an inch long; stamens very unequally inserted.

2. GILIA. Ruiz & Pav.

Corolla funnel-form, salver-form, or sometimes short-campanulate or rotate, regular. Stamens equally inserted (but sometimes with unequal filaments), not declined. Leaves various.

* *All of the leaves opposite, at least on the main stems, sessile and palmately parted or rarely entire. (Seeds mucilaginous in water.)*

Corolla from short funnel-form to almost rotate; the lobes obovate; filaments slender; anthers oval. Low or slender, loosely and mostly small flowered annuals; the leaves with filiform or setaceous divisions, appearing as if whorled. In ours, the flowers on filiform pedicels, loosely paniculate. § 1. *Dactylophyllum.*

Corolla salver-form, but the tube shorter than the calyx, the broad cuneate-obovate lobes slightly crenulate, strongly convolute in the bud; stamens inserted low on the corolla tube, included; erect, smooth; leaves entire or 3 5-divided. § 2. *Linanthus.*

Corolla salver-form, with usually a filiform elongated tube, and the throat sometimes abruptly dilated; stamens inserted in the throat; anthers short. Erect annuals, with leaves as in the last, and the flowers in a terminal capitate cluster. § 3. *Leptosiphon.*

* * *All the leaves alternate and palmately parted.*

Corolla similar to § 3. Stems woody; leaves much fascicled in the axils, 3 7-parted, rigid; flowers sessile, solitary or few at the ends of short branches. § 4. *Leptodactylon.*

* * * *All, or all but the lowest leaves alternate and pinnately compound, cleft or toothed, or rarely entire.*

Flowers capitate-glomerate or densely clustered, leafy-bracted; bracts and calyx-lobes often laciniate, rigid-accrose or spinulose-tipped. Corolla slender tubular-funnelform, with small oblong lobes; cells of the ovary and stigmas sometimes only 2. Annuals, mostly viscid-pubescent, never white-woolly, with once or twice pinnatifid leaves, their lobes commonly pungent; the bracts sometimes palmately cleft. § 5. *Navarretia.*

Flowers, inflorescence, etc., nearly as in § 5; but the anthers always exserted; corolla salver-form, more conspicuous; plants all white-woolly, not viscid. § 6. *Hugelia.*

Flowers capitate-glomerate, or panicled, or scattered, usually bractless; corolla (blue, purple or violet) from funnel-form to campanulate or almost rotate; stamens included or not surpassing the corolla lobes; leaves mostly pinnately incised. § 7. *Eugilia.*

§ 1. *Dactylophyllum.* Benth.

1. **G. liniflora**, Benth. From a few inches to over a foot high; leaves with nearly filiform divisions an inch long; corolla white, rotate, when fully open, 10 to 6 lines across, 5-parted down to the very short tube.

Var. **pharnaceoides**, Gr., is similar but smaller; the (sometimes pinkish) corolla half an inch across, or less.

2. **G. pusilla**, Benth. Small, 2 to 6 inches high; leaves less than half an inch long, shorter than the scattered pedicels; corolla nearly white, or purplish with a yellow throat, 1½ to 2 lines long, little exceeding the calyx.

Var. **Californica**, Gr., has a corolla 3 lines long, twice the length of the calyx; the throat often brownish. The most frequent form.

3. **G. Bolanderi**, Gr. Very like the last, but the tube of the blue or purple tinged corolla longer and narrower (3 or 4 lines long).

4. **G. aurea**, Nutt. Diffuse, 2 to 4 inches high; divisions of roughish leaves narrowly linear, 3 lines long; peduncles shorter or but little longer than the flowers; corolla usually yellow, short, funnel-form half an inch or less across; the roundish-obovate lobes about the length of the obconical throat and the short proper tube.

POLEMONIACEÆ. (GILIA FAMILY.)

Var. decora, Gr. Corolla white or pale violet, with or without a brown-purple throat; peduncles longer.

§ 2. *Linanthus*, Endl.

5. G. dichotoma, Benth. A span to a foot high, remotely leaved; flowers nearly sessile in the forks, or terminating the branches; calyx-tube white scarious; the teeth green; corolla white; the lobes from half to nearly an inch long; the tube sometimes purplish.

§ 3. *Leptosiphon*, Endl.

6. G. densiflora, Benth. A span to 2 ft. high; leaves in somewhat distant apparent whorls; tube of the white or rose-purple corolla about equaling the villous-hirsute bracts and calyx; its lobes nearly half an inch long, obovate.

7. G. androsacea, Steud. Erect or spreading, 3 to 12 inches high; corolla lilac, rose, pink or almost white, with a yellow or dark throat; its tube about an inch long.

Var. rosacea, Gr., is a dwarf tufted form with many rose-red flowers.

8. G. micrantha, Steud. Slender, about a span high; tube of the corolla very slender, 9 to 18 lines long; the lobes 2 or 3 lines long, from yellow to cream color and pale purple, or whitish.

9. G. tenella, Benth. Low and mostly depressed; tube of the corolla 6 to 9 lines long, the rose colored or pink lobes barely a line and a half long, the throat yellow; bracts and leaves hispidulous-ciliate.

10. G. ciliata, Benth. More rigid and hirsute than the preceding, a span to a foot high; tube of the rose-colored or purple, or in age whitish corolla, little if at all exserted beyond the very hirsute or hispid-ciliate bracts and subtending leaves, the lobes only a line and a half long.

§ 4. *Leptodactylon*, Hook & Arn.

11. G. Californica, Benth. Two or three feet high, with spreading rigid branches; corolla rose-color or lilac, an inch and a half in diameter.

§ 5. *Navarretia*, Gr.

* *Stamens included in the throat of the corolla.*

12. G. squarrosa, Hook & Arn. Rigid, rather stout, becoming much branched, very glandular-viscid, fetid with the odor of a skunk; upper leaves and bracts spinescent; corolla blue, 4 or 5 lines long.

* * *Stamens more or less exserted; corolla slender, 3 to 5 lines long. Leaves twice pinnatifid.*

13. G. cotulæfolia Steud. Rather stout and rigid, a foot or much less in height; villous pubescent and minutely glandular; upper bracts spinescent; tube of the violet or whitish corolla hardly longer than the calyx; capsule usually 1-seeded. Exhales the odor of *Anthemis cotula* (MAYWEED).

14. G. intertexta, Steud. At length diffusely much branched, a span high, neither

POLEMONIACEÆ. (GILIA FAMILY.) 79

viscid nor glandular; stems retrorsely pubescent; leaves mainly smooth, scarcely bipinnatifid; base of the bracts and tube of the calyx densely white-villous; corolla white.
15. G. leucocephala, Gr. A span high, rather slender, loosely branched, smooth, except a little woolliness at the top; leaves soft; bracts hardly pungent; heads dense; corolla white, longer than the calyx.

* * * *Stamens exserted; leaves only once pinnatifid, rigid, linear; corolla violet or purple, barely half an inch long, about twice the length of the pungent calyx-lobes.*
16. G. viscidula, Gr. A span high or less, at length much branched, viscid-pubescent; bracts palmately cleft.
17. G. atractyloides, Steud. Much more rigid than the last; leaves broader, the floral ovate, all with subulate spiny lobes; few flowered.

§ 6. *Hugelia*, Benth.
* *Root perennial; stems woody at the base.*
18. G. densifolia, Benth. A foot or two high; stems leafy, leaves linear, rigid, the short lobes subulate; flowers numerous in a compact head; corolla over half an inch long, violet blue, exceeding the calyx, the lobes 3 lines long; anthers sagittate.

* * *Root annual, stems slender, a foot or less in height; leaves and their few (if any) divisions filiform.*
19. G. virgata, Steud. Tube of the blue corolla longer than the calyx; anthers sagittate.
Var. floribunda, Gr. Low and rather stout; even the upper leaves pinnately 3-7-parted; the numerous heads and flowers as large as *G. densifolia*.

§ 7. *Eugilia*, Benth.
* *Flowers numerous in dense head-like clusters on long naked peduncles; stems erect; stamens inserted in the very sinuses of the short and broad corolla; leaves twice or thrice pinnately dissected into linear divisions.*
20. G. capitata, Dougl. Mostly smooth; stem slender, loosely branched above, a foot or two high; lobes of the light blue (rarely white) corolla narrowly oblong, 2 lines long.
21. G. achilleæfolia, Benth. Stouter and lower than the last, often glandular; the capitate clusters and flowers larger; calyx woolly; lobes of the deeper blue corolla broad.

* * *Flowers in small, rather loose clusters, or scattered in an open panicle.*
22. G. multicaulis, Benth. A span to a foot high, simple in early plants, loosely branched in later; flowers few in a cluster terminating the slender naked peduncles, almost sessile; the violet corolla 4 lines long, tube shorter than the viscid calyx; throat funnel-form; capsule ovoid.

80   HYDROPHYLLACEÆ. (WATERLEAF FAMILY.)

Var. tenera, Gr., is a depauperate form; frequently the peduncles only 1-flowered.
23. G. tricolor, Benth. A span to a foot or two high, in age diffusely branched; flowers few, in loose, rather short-peduncled clusters; corolla with a very short proper tube and an ample campanulate throat which is pale yellow or orange below, dark purple above; the lilac or violet lobes longer than the stamens.
24. G. inconspicua, Dougl. A span to a foot high, somewhat viscid or glandular; corolla violet-purple or bluish, twice or thrice the length of the calyx, but small, the lobes only a line long. It passes by gradation into
Var. sinuata, Gr., with the tube of the corolla more slender and exserted and the lobes often 2 lines long.

3. POLEMONIUM. Tourn.

Flowers as in *Gilia*, § *Eugilia*, but the corolla short and broad, the stamens somewhat declined, the filaments hairy appendaged at the base. Calyx herbaceous, its divisions and those of the pinnate leaves pointless.
1. P. cæruleum, L. (GREEK VALERIAN.) Smooth or viscid-pubescent, 2 or 3 ft. high, leafy, usually bearing numerous flowers; corolla an inch or more across. bright blue varying to white; stamens and style exserted.

ORDER 39. HYDROPHYLLACEÆ.

Inflorescence usually scorpioid; flowers perfect, regular, 5-androus, the two styles distinct at least at the apex; stigmas terminal, small, capitate. Only in *Romanzoffia* are the stigmas as well as the styles united. Ovary commonly hispid or hirsute, at least at the top.—Mostly herbs, with alternate or rarely opposite leaves and no stipules.
Tribe 1. HYDROPHYLLEÆ. Ovary and capsule 1-celled. Style 2-cleft. Corolla almost always convolute in the bud. Herbs.

Flowers solitary or loosely racemose.
Calyx with reflexed appendages.....................................Nemophila. 1
Calyx naked at the sinuses...Ellisia. 2

Tribe 2. PHACELIEÆ. Ovary 1-2-celled. Style 1-2-cleft. Corolla imbricated in the bud. Calyx naked at the sinuses. Herbs.

Corolla not yellow, deciduous.....................................Phacelia. 3
Corolla yellow, persistent.. Emenanthe. 4
Style and stigma entire...Romanzoffia. 5

Tribe 3. NAMEÆ. Ovary, capsule, dehiscence, etc., nearly of *Phaceliæ*. Styles distinct to the base, stigmas capitate.
Low shrubs...Eriodictyon. 6

HYDROPHYLLACEÆ. (WATERLEAF FAMILY.) 81

1. NEMOPHILA, Nutt.

Calyx 5-parted. Corolla rotate-campanulate, deeply 5-lobed, the throat appendaged with 10 internal plates or scales.—Tender herbs with diffuse and procumbent stems, and pinnately lobed or divided leaves, more or less hirsute.

* *Leaves mostly alternate; stems long and weak, beset with stiff reflexed bristles.*

1. **N. aurita**, Lindl. Leaves large, with auriculate dilated and clasping base or winged petiole deeply pinnatifid into 5 to 9 retrorse lobes; corolla violet, 5 to 12 lines in diameter.

* * *Leaves opposite not auricled at base.*

2. **N. maculata**, Benth. Leaves lyrately pinnatifid into 5 to 9 short lobes, or the uppermost only 3-lobed; corolla white, with a violet spot at the top of each lobe, over an inch across.
3. **N. insignis**, Dougl. Leaves similar to the last; corolla bright blue, its scales short and roundish, partly free.
4. **N. Menziesii**, Hook & Arn. Leaves less divided than the last; corolla from light blue to white and sprinkled with dots toward the center, its scales narrow and adherent by one edge.

* * * *Upper leaves often alternate, mostly longer than the peduncles, and slender-petioled, many only 3–5-lobed, one-sided.*

5. **N. parviflora**, Dougl. Slender and weak; corolla 2 to 5 lines across, light blue or white.

2. ELLISIA, L.

Calyx 5-parted. Corolla campanulate, short in proportion to the calyx; scales minute or obsolete. Stamens and style not exserted.

1. **E. chrysanthemifolia**, Benth. Stem 1 or 2 ft. high, erect, branched; leaves dissected into very many small and short divisions; flowers, small, white; capsule remarkable, viz.: the mostly four ordinary rough seeds enclosed between the placentæ, while, between each placenta and the valve which it lines, is hidden a single thin, meniscoidal, smooth seed.

3. PHACELIA, Juss.

Calyx deeply 5-parted, the divisions usually narrow and similar; corolla from almost rotate to narrow-funnelform; commonly with appendages upon the inside of the tube in the form of 10 vertical plates, approximate in pairs between the bases of the filaments, or adnate to the filaments, one on each side. Stamens equally inserted low down or at the base of the corolla. Herbs, mostly hirsute or hispid and branched from the base; with simple or compound alternate leaves, or the lower opposite and more or less scorpioid inflorescence. Corolla never yellow except in the throat. Ovules and seeds 4 in all except the last species.

6

82   HYDROPHYLLACEÆ.   (WATERLEAF FAMILY.)

* *Leaves simple and entire, or with a pair or two of similar and smaller leaflets or lobes.*

1. **P. circinata**, Jacq. f. A span to a foot or two high from a stout root, hispid and the foliage strigose, either green, grayish or canescent, with a soft pubescence; leaves from lanceolate to ovate, acute, the lower tapering into a petiole and some bearing lateral leaflets; inflorescence in dense scorpioid hispid spikes, crowded; corolla dull or bluish white; filaments much exserted.—A very variable species; usually many stems from one root; some with large entire, ovate green leaves only.

2. **P. Breweri**, Gr. Foliage and habit similar to the last, but smaller and more slender, from an annual root; leaves seldom an inch long, many of them 3-5-parted, the lanceolate lateral lobes ascending; corolla smaller (scarcely 3 lines long), blue or violet; filaments not exserted.

* * *Leaves simple, rounded, cordate, lobed and serrate.*

3. **P. malvæfolia**, Cham. Stout, loosely branching, hispid with stinging hairs; leaves 2 inches or more in diameter; spikes solitary, or in pairs; corolla 3 to 6 lines long, dull white or bluish; stamens much exserted.

* * * *Leaves once to thrice pinnatifid or pinnately compound, oblong in general outline. Calyx bristly hispid, its lobes not rarely unequal. Annuals, the species difficult to discriminate.*

4. **P. tanacetifolia**, Benth. Erect, 1 to 3 ft. high, roughish, hirsute or hispid; leaves 9-17-divided in narrow once or twice pinnately parted or cleft divisions, all sessile or nearly so; the scorpioid spikes clustered; the short pedicels erect or ascending; corolla usually of a dirty mottled white or bluish; stamens and style much exserted; calyx lobes not twice the length of the capsule.

5. **P. ramosissima**, Dougl. Straggling, somewhat viscid above; leaves pinnately 5-7-divided or parted into linear pinnatifid-incised divisions; the short pedicels soon horizontal; stamens and style moderately exserted; calyx lobes more than twice the length of the globular capsule; flowers bluish.

6. **P. ciliata**, Benth. A span to a foot high; leaves rarely divided but incised or cleft and toothed; spikes simple or in pairs; stamens usually not surpassing the open corolla; calyx lobes ciliate with glandular bristles; corolla blue.

* * * * *Leaves entire, or the lower 1-2-lobed, not cordate, the veins parallel or converging, as in P. circinata; no glandular pubescence; calyx with long hairs; seeds more than 4.*

7. **P. divaricata**, Gr. Diffusely spreading, a span or more in height; leaves ovate or oblong; style 2-cleft at the apex only; corolla violet, about 10 lines in diameter.

4. EMMENANTHE, Benth.

Distinguished from *Phacelia* by the persistent yellow or cream-colored corolla.

1. **E. penduliflora**, Benth. A span to a foot high; somewhat viscid; leaves pinnatifid; pedicels filiform, about half an inch long, equaling the nodding corolla.

5. ROMANZOFFIA, Cham.

Stamens unequal; style filiform. Low perennial herbs, with the aspect of saxifrages; the leaves mainly radical, round-cordate, or reniform, crenately 7–11-lobed, long petioled.

1. **R. Sitchensis**, Bong. Scapes weak, a span long, bearing several pink or purple, varying to white flowers; corolla veiny.

6. ERIODICTYON, Benth.

Calyx deeply 5-parted. Corolla funnel-form to salver-form. Stamens included.—Low shrubs; the leaves alternate, of rigid coriaceous texture, the finely reticulated veinlets conspicuous on a fine woolly ground, at least underneath, their margins beset with rigid teeth.

1. **E. glutinosum**, Benth. (MOUNTAIN BALM, or YERBA SANTA.) Smooth, glutinous with a resinous exudation, 3 to 5 ft. high; leaves lanceolate, 3 to 6 inches long; cymes in a naked panicle; corolla tubular, funnel-form, violet or nearly white, half an inch long.

E. tomentosum, Benth., grows farther down the coast. It is larger with smaller almost salver-form flowers; densely villous.

ORDER 40. BORRAGINACEÆ.

Mostly roughly pubescent herbs, with alternate entire leaves without stipules, scorpioid inflorescence, and perfectly regular 5-androus flowers; the ovary of 4 lobes or divisions around a central style, ripening into seed-like nutlets. Calyx free, 5-parted or 5-cleft, persistent. Corolla with a 5-lobed limb, commonly imbricated in the bud. Stamens distinct, inserted in the tube or throat of the corolla alternate with its lobes. The one-sided and coiled apparent spikes or racemes straighten as the blossoms develop. All our species except the first belong to the true Borrage Tribe.

* *Fruit not prickly.*

Corolla with plaited sinuses; stigma sessile..........................Heliotropium. 1
Corolla yellow. Bristly-hispid plants...............................Amsinckia. 2
Corolla white..Eritrichium. 3

* * *The nutlets prickly, bur-like.*

Flowers sky-blue (rarely white) in bracteate racemes............Echinospermum. 4
Flower purple, blue and violet in a peduncled raceme..............Cynoglossum. 5
Flowers minute; nutlets winged, or boat-shaped.....................Pectocarya. 6

BORRAGINACEÆ. (BORRAGE FAMILY.)

1. HELIOTROPIUM, Tourn.

Corolla with plaited sinuses. Filaments short or none; anthers connivent and sometimes cohering. Style entire or none; stigma a fleshy ring or the edge of a peltate or umbrella-shaped disk. Fruit dry, splitting into 4 nutlets.

1. **H. Curassavicum**, L. A smooth and somewhat glaucous succulent herb with spreading or prostrate stems; leaves oblanceolate, an inch or two long; flowers crowded, white or blue; stigma sessile, flat-topped. Blackens in drying.

2. AMSINCKIA, Lehm.

Corolla salver-form, or somewhat funnel-form, more or less plaited in the bud at the sinuses, with the tube exceeding the calyx, lobes rounded. Filaments short. Style filiform; stigma capitate-2-lobed. Nutlets ovate-triangular. Hispid annuals with oblong-ovate to linear leaves, and yellow flowers in at length loose scorpioid spikes or racemes, without bracts, except sometimes the lowest.

* *Nutlets rough, the back convex.*

1. **A. spectabilis**, Fisch. & Mey. Erect, a span to a foot high; leaves mostly linear; tube of the bright orange-yellow corolla, two or three times the length of the linear, rusty-hispid calyx, nearly half an inch long; the throat enlarged, and the expanded limb a third to half an inch in diameter.

2. **A. intermedia**, Fisch. & Mey. Erect, usually a foot or two high; leaves linear or only the lower lanceolate; corolla bright yellow, 3 or 4 lines long; its tube a little surpassing the calyx-lobes; the limb 2 or 3 lines in diameter.

3. **A. lycopsoides**, Lehm. Loosely branched, soon spreading, sometimes decumbent, sparsely hispid with bristles, which on the leaves have conspicuous pustulate bases; leaves from lanceolate to ovate, the margins usually undulate; upper flowers mostly bractless; corolla light yellow, about 4 lines long; the throat little enlarged; the limb 2 or 3 lines in diameter. Passes into

Var. **bracteosa**, Gr., a smaller-flowered decumbent form, with most of the flowers bracteate.

* * *Nutlets nearly flat on the back, coarsely granulate.*

4. **A. tessellata**, Gr. About a foot high, rather stout, coarsely hispid, the bristles of the calyx rusty; corolla orange-yellow, 3 or 4 lines long, the throat plaited, the tube rather longer than the obtuse calyx-lobes; nutlets broadly ovate, thickly covered with warty granulations closely fitting like the blocks of a pavement.

* * * *Nutlets at maturity, whitish, smooth and polished.*

5. **A. vernicosa**, Hook & Arn. Sparsely bristly; leaves linear to ovate-lanceolate; corolla light yellow, 4 or 5 lines long, and the limb narrow; nutlets shaped like a grain of buckwheat.

BORRAGINACEÆ. (BORRAGE FAMILY.)

Var. grandiflora, Gr. Robust, more hispid and large flowered, the limbs broader; calyx lobes often combined, so as to appear as 3 or 4.

3. ERITRICHIUM, Schr.

Most obviously distinguished from *Amsinckia* and the nearer *Echinospermum* by its usually smaller white flowers, with shorter corolla tube. The species difficult of determination.

1. **E. Californicum**, DC. The slender stems decumbent, a span or more long; the leaves narrowly linear; stems flowering from near the base; flowers almost sessile, mostly with leaves or bracts, at length scattered; the corolla only a line long; calyx open in fruit. Passes into

Var. **subglochidiatum**, Gr. Slightly succulent; lower leaves inclined to spatulate, nutlets somewhat barbed. Wet ground.

2. **E. Scouleri**, A. DC. Slender, erect a span to a foot high; leaves narrowly linear (1 or 2 inches long); flowers in geminate or sometimes paniculate slender naked spikes, most of them bractless; pedicels not more than a line long; calyx erect in fruit; corolla surpassing the calyx, the limb almost rotate, 2 to 5 lines in diameter.—Seems to pass into the next.

3. **E. Chorisianum**, DC. At first erect, soon spreading or decumbent; larger leaves, 2 to 4 inches long; flowers in lax, usually solitary racemes, many of them leafy-bracted pedicels sometimes filiform and 2 to 9 lines long; corolla more funnel-form, its limb 3 to 5 lines in diameter.—This may be a wet ground form of the last, which grows on dry ground.

4. **E. fulvum**, A. DC. A span to a foot high, slender branched from a leafy base, pubescent; leaves linear, or the lower lanceolate or spatulate; spikes at maturity nearly filiform, bracteate only at the base; calyx, etc., densely clothed with rusty or fulvous hairs; calyx deciduous, only the lower part remaining under the fruit; corolla limb 2 lines across.

5. **E. canescens**, Gr. Stouter and larger than the last; the pubescence whitish, not rusty; leaves linear; calyx hardly deciduous.

6. **E. oxycaryum**, Gr. May be known by the solitary ovate-acuminate, smooth, shining nutlet enclosed in the persistent bur-like calyx; corolla 2 lines wide.

4. ECHINOSPERMUM, Swartz.

Calyx lobes spreading or reflexed in fruit. Corolla short, salver-form, and with conspicuous arching crests at the throat. Short filaments, style, etc., as in *Eritrichium*. Nutlets with barbed prickles.

1. **E. floribundum**, Lehm. Rather strict, 2 ft. or more high, or sometimes smaller; leaves from oblong to linear-lanceolate; racemes numerous, usually geminate; the tri-

86   CONVOLVULACEÆ. (MORNING GLORY FAMILY.)

angular nutlets armed with prickles on the margins; limb of the rotate corolla 2 to 5 lines in diameter, blue, rarely white.

5. CYNOGLOSSUM, Tourn.

Chiefly distinguished from the preceding by the broad large leaves, the bractless racemes and the nutlets clothed over the whole back with stout barbed prickles.

1. **C. grande**, Dougl. About 2 ft. high, pubescence soft; radical and lower stem leaves ovate oblong, usually rounded or cordate at the base, long petioled; panicled racemes or cymes small, on a long naked terminal peduncle; corolla tube exceeding the calyx; its limb blue to violet, with usually purple crests; 3 to 5 lines wide.

6. PECTOCARYA, DC.

Structure of the minute white flowers similar to the preceding; nutlets widely spreading in pairs, horizontal, oblong or almost linear, surrounded by an incurved wing-like border which is toothed, the apex beset with hooked bristles.

1. **P. penicillata**, A. DC. Very slender, diffusely branching, spreading, with narrow linear leaves, and small flowers scattered the whole length of the stem, on very short pedicels; nutlets only a line long.

ORDER 41. CONVOLVULACEÆ.

Herbs, usually twining or trailing, with alternate leaves (or scales) and regular perfect flowers; the stamens as many as the lobes or angles of the corolla and alternate with them (5, rarely 4); the free persistent calyx of mostly distinct imbricated sepals; ovary 2-3-celled; capsules generally globular; seeds 1 to 4. Inflorescence axillary.

Corolla plaited in the bud; style single............................ **Convolvulus.** 1
Corolla 5-cleft; styles 2... **Cressa.** 2
Twining parasites, leafless, yellowish................................ **Cuscuta.** 3

1. CONVOLVULUS, L.

Corolla campanulate or short and open funnel-form, with a 5-angulate or obscurely 5-lobed border, deeply plaited down the sinuses in the bud. Stamens included. Style filiform; stigmas 2, in ours flat, from linear to oval.

* *A pair of bracts close to the calyx, enveloping it.*

1. **C. Soldanella**, L. Maritime, low, smooth; stems a foot or less in length, trailing; leaves reniform entire or obscurely angulate-lobed, an inch or two broad, long petioled; corolla pink, purplish, or nearly white.

CONVOLVULACEÆ. (MORNING GLORY FAMILY.)

2. **C. occidentalis,** Gr. Mostly smooth; stems twining several feet high; leaves from broadly ovate-triangular with a deep and narrow basal sinus to narrowly lanceolate-hastate; the posterior lobes often 1-2-toothed; peduncle elongated, not rarely 2-flowered within the bracts; these ovate or rarely oblong, commonly surpassing the enclosed calyx; corolla white or pinkish, 1 to 1½ inches broad; stigmas linear.

3. **C. Californicus,** Choi. Minutely and rather densely pubescent, a span or less high, or with trailing stems a foot long; leaves from ovate or obovate and obscurely hastate to triangular-hastate, the basal lobes sometimes 1-2-toothed, long-petioled; peduncles shorter than the petiole; bracts oblong or oval, about equaling the sepals, or shorter; corolla white, cream-color or flesh-color, 1½ to 2 inches long.

4. **C. villosus,** Gr. Densely silky-villous or woolly; corolla cream colored, an inch long.

* * *No calyx-like bracts; sometimes a pair of leaves close under the flower or a pair of bracts at some distance below it.*

5. **C. luteolus,** Gr. Stems twining several feet long; leaves triangular-hastate or sagittate, the basal lobes sometimes 2-lobed; peduncles bearing a pair of linear or lanceolate entire bracts, a little below the flower; a second flower occasionally from the axil of one of them; corolla pale yellow or purplish, an inch or more in length; stigmas linear.

2. CRESSA, L.

Corolla deeply 5-cleft; the oblong or ovate lobes more than half the length of the somewhat campanulate tube. Stamens and the 2 distinct styles exserted. Stigmas capitate.

1. **C. Cretica,** L. A span or two high, silky-villous and hoary; leaves very numerous, 2 to 4 lines long, almost sessile; flowers sessile or nearly so in the upper axils; corolla 2 or 3 lines long, white.—On saline or alkaline soil.

3. CUSCUTA, Tourn. DODDER.

Calyx 5-4-cleft or parted. Corolla campanulate or short-tubular, the spreading limb 5-4-parted. Styles in our species 2, distinct. Seeds germinating in the soil, but the thread-like, branching, leafless, yellowish or reddish twining stems becoming parasitic on the bark of herbs or small shrubs; being attached by means of suckers. Flowers small, cymose or densely clustered, white or whitish.

* *Capsule depressed-globose.*

1. **C. Californica,** Choisy. Flowers pedicelled in loose few-flowered cymes; lobes of the calyx acute; lobes of the corolla lanceolate-subulate, delicate white; no scales below the stamens.

Var. **breviflora,** Engel. Flowers scarcely over a line long; calyx lobes equaling the corolla-tube.

SOLANACEÆ. (POTATO FAMILY.)

Var. longiloba, Engel. Flowers 1½ to 2½ lines long; calyx-lobes often with recurved tips; capsule mostly only 1-seeded, enveloped by the withered corolla.

* * *Capsule pointed, capped or enveloped by the withered corolla.*

2. C. salina, Engel. Flowers 1½ to 2½ lines long delicate white; corolla lobes often overlapping, denticulate; capsule surrounded but not capped by the corolla, usually 1-seeded.—Growing in saline marshes, usually on *Salicornia*.

3. C. subinclusa, Dur. & Hilg. Flowers sessile or nearly so (at length in large clusters), 2½ to 4 lines long; lobes of the corolla short, the tube somewhat urn-shaped, only partly covered by the fleshy, usually reddish calyx.—The most common species growing on coarse herbs and shrubs.

Order 42. **SOLANACEÆ**.

Herbs or shrubs, with alternate leaves and no stipules, regular 5-merous flowers on bractless pedicels, a single style and a 2-celled ovary; the fruit a many-seeded berry or capsule.

Corolla rotate; fruit a berry .... Solanum. 1
Corolla funnel-form; capsule large, spiny .... Datura. 2
Corolla funnel-form; capsule smooth .... Nicotiana. 3

1. **SOLANUM**, Tourn.

Lobes of the corolla valvate in the bud. Filaments short; anthers usually conniving. Style elongated.

* *Corolla small, white; deeply 5-cleft.*

1. S. nigrum, L. (Black Nightshade.) Widely branching; leaves usually ovate and sinuate toothed; flowers in umbellate clusters; berries black. Variable.

Var. Douglasii, Gr. Leaves apt to be coarsely toothed; flowers sometimes half an inch broad.

* * *Corolla large, blue, 5-angled.*

2. S. umbelliferum, Esch. Somewhat shrubby; flowers in umbel-like clusters, violet-blue to rarely white, about 9 lines broad.—A variable species similar to *S. Xanti* (which is less shrubby and has larger flowers), a common species farther south.

2. **DATURA**, L. Stramonium.

Calyx prismatic, partly deciduous. Corolla with ample 5-pointed limb. Style long; stigma 2-lipped. Capsule spiny.

1. D. Stramonium, L. Smooth, green; corolla white, about 3 inches long; capsule beset with short stout prickles, the lower shorter.

SCROPHULARIACEÆ. (FIGWORT FAMILY.) 89

2. **D. Tatula**, L. Stem reddish-purple; corolla pale violet; prickles about equal.
3. **D. quercifolia**, HBK. Green; corolla violet-tinged; prickles flattened, unequal, some an inch long.—Lower Russian River.

3. NICOTIANA. Tourn. TOBACCO.

Calyx campanulate or oblong, persistent. Corolla commonly funnel-form, the limb plaited. Style long; stigma capitate, somewhat 2-lobed.—Very viscid herbs.
1. **N. rustica**, L. Leaves petioled, ovate, or the lower slightly cordate; corolla short and broad, dull-white, less than an inch long.
2. **N. Bigelovii**, Wat. Leaves oblong or oblong-lanceolate, only the lower ones petioled, these scarcely exceeding 6 inches long; corolla nearly salver-form with tube 1½ inches long, the limb an inch or more wide, its lobes acute.

ORDER 43. SCROPHULARIACEÆ.

A corolla more or less bilabiate with the lobes imbricated in the bud; didynamous or diandrous stamens; a single style and a 2-celled ovary and capsule marks this large order. In *Pentstemon* there is a fifth rudimentary stamen. *Verbascum* has five perfect stamens.

Two species of *Verbascum* (*Mullein*) are found in the State, but probably not within our limits: *V. Thapsus*, L., with woolly decurrent leaves and *V. virgatum*, Withe., distinguished by nearly smooth not decurrent leaves and violet bearded filaments.

* *Leaves mostly alternate; corolla personate.*
Corolla spurred at base...Linaria. 1
Corolla gibbous at base...Antirrhinum. 2

* * *Leaves opposite or whorled.*
Corolla erect, the anterior lobe reflexed, the other 4 erect, a scale in the throat on
    the upper side...Scrophularia. 3
Corolla declined, the middle lower lobe infolding the stamens and style...Collinsia. 4
Carolla with a fifth sterile filament on the upper side................Pentstemon. 5
Stigma 2-lipped or disk-like..Mimulus. 6

* * * *Corolla rotate or short-campanulate.*
Calyx 5-toothed; corolla campanulate..................................Limosella. 7
Calyx 4-parted; corolla 4-lobed, rotate................................Veronica. 8

* * * * *Corolla tubular; the upper lip erect or incurved, laterally compressed, usually enclosing the ascending stamens.*
Corolla narrow with almost obsolete lower lip.........................Castilleia. 9
Corolla with saccate lower lip of 3 lobes.............................Orthocarpus. 10

SCROPHULARIACEÆ. (FIGWORT FAMILY.)

Lips of corolla, both short; the lower 3-crenulate.................**Cordylanthus. 11**
Upper lip of the corolla arched; many large radical leaves...........**Pedicularis. 12**

1. LINARIA, Tourn.

Calyx 5-parted. Corolla with the throat nearly closed; the base in front (below) prolonged into a spur.

1. **L. Canadensis,** Dum. (TOAD FLAX.) Smooth; leaves linear, alternate on the erect flowering stems, but smaller and broader ones often opposite or whorled on the procumbent shoots; flowers blue in a terminal raceme.

2. ANTIRRHINUM, Tourn. SNAPDRAGON.

Like *Linaria*, except that the corolla has a saccate protuberance instead of a spur. In ours the upper lip is spreading and the lower lobes deflexed.

1. **A. glandulosum,** Lindl. Glandular and viscid; leaves lanceolate, mostly sessile; flowers in a dense spike or raceme, half an inch or more long, pink with yellowish palate.

2. **A. vagans,** Gr. Very diffuse, often glandular, branchlets frequently prehensile; leaves short, lanceolate to ovate; flowers scattered, purplish blue, half an inch long.
Var. **Bolanderi,** Gr. Has broader and thinner leaves, those on the prehensile branchlets orbicular.

3. **A. Breweri,** Gr. Has smaller flowers, only 3 lines long; style strongly deflexed.

3. SCROPHULARIA, Tourn. FIGWORT.

Calyx deeply 5-cleft, the lobes broad. Corolla short, with an oblong tube unequally 5-lobed, 4 erect, the two upper the longer. Stamens 4, inserted in pairs, low down on the corolla tube, a rudiment of the fifth stamen in the form of a scale above. Coarse herbs, with inconspicuous flowers.

1. **S. Californica,** Cham. Nearly smooth, 2 to 6 ft. high, with deltoid or truncate-ovate doubly toothed opposite leaves; flowers small greenish or lurid red (rarely yellow) in a terminal thyrsus.

4. COLLINSIA, Nutt.

Calyx deeply 5-cleft. Corolla with the tube gibbous or saccate on the upper side, commonly declined, conspicuously bilabiate; the upper lip 2-cleft, and its lobes recurving; the lower 3-lobed and larger, its side lobes pendulous-spreading, the middle one folded into a keel-shaped sac and including the declined stamens and style. Stamens in pairs, with long filaments, anthers round-reniform. A gland at the base of the corolla on the upper side answers to the fifth stamen.—Beautiful annuals with simple opposite or whorled leaves, all but the lower sessile; pedicels solitary or whorled in the axils of leaves which diminish to small bracts above.

SCROPHULARIACEÆ. (FIGWORT FAMILY.) 91

* *Flowers short-pediceled or nearly sessile, verticillate.*

1. **C. bicolor**, Benth. A foot or more high; leaves oblong-lanceolate, the upper usually ovate-lanceolate and sessile by a nervose veined base; pedicels shorter than the acute lobes of the calyx; the lower lip or the corolla violet or rose-purple and the upper paler to nearly white; the saccate throat very oblique to the true tube, fully as broad as long; gland short.—The most showy species, with flowers nearly an inch long.

2. **C. tinctoria**, Hartw. Foliage, etc., like the preceding; generally more viscid-pubescent; flowers almost sessile; corolla yellowish, cream-color, or white, usually with purple dots or lines; upper lip very short.—East side of Sacramento Valley.

3. **C. bartsiæfolia**, Benth. Puberulent and somewhat glandular; leaves from ovate-oblong to linear; flower-whorls 2 to 5, rarely only one; the lateral lobes of the lower lip emarginate or obcordate; gland elongated. Flowers nearly as large as the preceding, purplish, pale violet, or whitish; upper lip with a transverse callosity at the origin of the limb.

4. **C. Greenei**, Gr. Upper lip of the violet purple corolla about half the length of the lower, crested below with a pair of callous teeth on each side connected by a ridge. Corolla 5 lines long.—Lake County.

* * *Flowers on slender pedicels, solitary or umbellate-whorled.*

6. **C. sparsiflora**, Fisch. & Mey. Slender; upper leaves linear-oblong or linear-lanceolate, merely opposite or the upper minute floral bracts in threes; pedicels solitary in the axils, longer or shorter than the flower which is 4 to 8 lines long; corolla mostly violet; the upper lip and the middle lobe of the lower commonly yellowish and purple-dotted; calyx usually purple-tinged.

7. **C. parviflora**, Dougl. Low, at length diffuse about a span high; the blue, or partly white flowers solitary or 2 to 5 in a whorl, 2 to 4 lines long; stigma cleft, gland capitate, short-stipitate.

5. PENTSTEMON, Mitch.

Calyx 5-parted. Corolla with a conspicuous mostly elongated or ventricose tube; the limb more or less bilabiate; upper lip 2-lobed; the lower 3-cleft, recurved or spreading.—The conspicuous sterile filament strongly marks the genus, remarkable for its many beautiful species.

1. **P. Menziesii**, Hook. Tufted at the woody base, a span to a foot high; leaves oval or ovate, a half to an inch long; corolla about an inch long, pink-red; anthers with the diverging cells long-woolly. Mt. St. Helena, *Mrs. M. L. Swett*.

2. **P. corymbosus**, Benth. A foot or two high, soft-pubescent or nearly smooth, leafy to the tip; corolla scarlet, an inch long; anthers smooth; steril filament, bearded down one side.

3. **P. breviflorus**, Lindl. 3 to 6 ft. high, with long, slender, flowering branches; corolla yellowish with flesh-color, striped within with pink, about half an inch long; the upper lip beset with long viscid hairs; sterile filament naked.

4. **P. Lemmoni**, Gr. Is smaller and may be distinguished from the last by its yellow bearded sterile filament.

5. **P. heterophyllus**, Lindl. Stems 1 to 5 ft. high from a woody base; leaves lanceolate or linear; corolla an inch or more in length, ventricose, rose-purple or pink changing to violet, an inch or more in length. Difficult to distinguish from the next.—Coast Range.

6. **P. azureus**, Benth. Usually smaller than the last; the larger corolla azure blue changing to violet; the base sometimes reddish; the expanded limb sometimes an inch broad.—Sierra Nevada.

6. MIMULUS, L.

Calyx mostly plicately 5-angled. Corolla funnel-form, with the included or rarely exserted tube bilabiately 5-lobed; the lobes roundish, more or less spreading or the upper turned back; a pair of ridges running down the lower side of the throat. The anthers often approximate in pairs, their cells divergent. The lobes of the stigma commonly petaloid-dilated or peltate-funnelform.—Flowers axillary on simple peduncles; commonly showy.

1. **M. tricolor**, Lindl. Stem, when beginning to flower, only a quarter of an inch high, at length 3 inches. Corolla about 1½ inches long, with a long exserted slender tube, a short funnelform throat, and similar nearly equal lobes; pink, with a crimson spot on the base of each lobe, a yellow stain along the lower lip. Leaves sessile.

2. **M. Douglasii**, Gr. Similar to the last; leaves contracted into a petiole; lower lip of the corolla much shorter than the erect upper one or even obsolete; the throat more ample. Stem from a ¼ to 6 inches high.

3. **M. glutinosus**, Wendl. A brittle-stemmed shrub, 2 to 6 ft. high, with thick glutinous-sticky leaves and mostly buff or salmon-colored flowers, but running into varieties with red, red-brown, or scarlet flowers.

4. **M. cardinalis**, Dougl. Villous, with viscid hairs; the large leaves ovate, the upper often connate; corolla frequently 2 inches long; the tube hardly exceeding the long calyx, the limb very oblique, scarlet.—Along water courses.

5. **M. luteus**, L. Mostly smooth, varying greatly in size from a foot to even 4 ft. high; leaves ovate oval or cordate; corolla deep yellow, usually spotted within, and the base of the lower lip blotched with brown-purple, from 1 to 2 inches long. Moist ground.

6. **M. inconspicuus**, Gr. Smooth, 2 to 7 inches high; the ovate or lanceolate leaves sessile, a half inch or less long; corolla 5 lines long, yellow or rose-color; calyx teeth very short.

7. **M. moschatus**, Dougl. (MUSK PLANT.) Very villous and usually musk-scented; stems spreading and creeping; flowers yellow.—Our form is chiefly

Var. **longiflorus**, Gr., with very clammy leaves and flowers an inch long, scarcely musky.

SCROPHULARIACEÆ. (FIGWORT FAMILY.) 93

8. **M. pilosus**, Wat. A span to a foot high, much branched, soft, villous and slightly viscid, many flowered from near the base; leaves lanceolate to narrowly oblong, sessile, entire; calyx tube not prismatic; corolla yellow, obscurely bilabiate, 3 or 4 lines long, usually a pair of brown-purple spots on the lower lobe.

7. LIMOSELLA, L. MUDWORT.

Calyx campanulate. Corolla rotate-campanulate, nearly regular. Style short; stigma thickish.—Diminutive annuals, with narrow fleshy leaves in clusters around the 1-flowered scapes. Flower small, white or purplish.

1. **L. aquatica**, L. An inch to a span high, growing in brackish mud or in fresh water.

8. VERONICA, L.

The lower lobe and sometimes the lateral ones of the rotate corolla sometimes smaller than the others. Stamens 2, one on each side of the upper lobe of the corolla. Capsules compressed. Flowers small (a line or two broad), in racemes or spikes, or solitary in the axils; blue, purplish, or white.

1. **V. Americana**, Schw. Stems a span to two feet long; leaves ovate or oblong, serrate, rather succulent, short-petioled, an inch or two long, opposite. Flowers in axillary racemes, bluish, with purple stripes. Common in damp places.

2. **V. peregrina**, L. A span or more high, all the upper leaves alternate, linear-oblong; flowers minute, in the axils of the leaves, and mostly narrow bracts; capsule obcordate.

9. CASTILLEIA, Mutis. PAINTED-CUP.

Calyx tubular, more or less cleft in front or behind, or both; the lobes 2 and lateral, or 4. Corolla tubular, laterally compressed, especially the long upper lip (galea); the lower lip very short or minute, 3-toothed, and somewhat saccate below the short teeth; the tube usually inclosed in the calyx. Stamens 4, inclosed in the galea; anthers 2-celled, the long cells unequal, the outer fixed by the middle, the inner ones smaller, pendulous. Style long; the capitate stigma sometimes 2-lobed. Herbs, sometimes woody at the base, with mostly alternate, sessile leaves, the floral ones or their tips, as well as the calyx lobes, commonly petaloid and colored red, yellow, or white. Flowers in terminal, simple, leafy spikes.

1. **C. affinis**, Hook. & Arn. Annual; a foot or two high; leaves narrowly lanceolate, entire; the upper floral bracts usually broader, the apex toothed, red; spike with scattered, frequently pedicellate flowers below; calyx red; an inch long, its front fissure hardly twice as deep as the back one, the narrow lobes acutely 2-cleft; corolla 1 to 1½ inches long, exserted so as to expose the callous lip; the galea about equal to the tube, yellowish or tipped with red.

2. **C. latifolia**, Hook. & Arn. Perennial (as are all the following); branching from

the base, 1 or 2 ft. high, villous-hirsute and viscid; leaves oval, obtuse, half an inch or more long, some above 3-5-lobed and red; calyx 2-cleft to the middle, the lobes entire or emarginate, almost equaling the corolla; corolla 8 lines long, the short teeth of the lip inflexed.

3. **C. parviflora**, Bong. A span to 2 ft. high, villous-hirsute above; leaves variously cleft into linear or lanceolate lobes, or sometimes the cauline are mainly entire and narrow; calyx lobes oblong and 2-cleft at the apex or to below the middle; corolla an inch or less long; only the upper part of the narrow galea exserted—A variable species. As in the preceding species, the bracts and calyx are usually colored red or crimson, but sometimes varying to yellow or even white.

4. **C. miniata**, Dougl. Commonly 2 ft. high, strict, often slender; leaves lanceolate or linear-lanceolate, almost always entire, the broad floral ones of the close spike sometimes incised or 3-cleft, usually bright red, rarely whitish; calyx lobes lanceolate, acutely 2-cleft; corolla over an inch long, exserted, exposing the short ovate teeth of the lip.

5. **C. foliolosa**, Hook. & Arn. Densely white-woolly, the matted hairs loosened with age; many-stemmed from a woody base; leaves narrowly linear, an inch or less long, crowded below and fascicled in the axils.

10. ORTHOCARPUS, Nutt.

Chiefly distinguished from *Castilleia* by the upper lip of the corolla (galea) which but little, if at all, surpasses the usually more conspicuous and inflated 1-3-saccate lower lip.

§ 1. CASTILLEIOIDES, Gr.—*Lower lip of the corolla simply or somewhat triply saccate, and bearing 3 conspicuous teeth; the galea broadish or narrow; stigma capitate; anthers all 2-celled; bracts with colored tips.*

* *Filaments smooth; galea straight or nearly so, naked, narrow; the lip moderately ventricose; its teeth erect.*

1. **O. attenuatus**, Gr. Slender, strict, a span or two high, mostly simple; leaves linear and attenuate, often with a pair of filiform lobes; spike slender; lower flowers scattered; bracts with slender lobes barely white-tipped; corolla narrow, half an inch long, white or whitish; narrow teeth of the purple-spotted lip nearly equaling the galea.

2. **O. densiflorus**, Benth. Erect or diffusely branched from the base 6 to 12 inches high; spike dense, many flowered, at length cylindrical, or lowest flowers rather distant; bracts 3-cleft, about equaling the flowers, their linear lobes purple and white; corolla from 8 to 12 lines long, the tips usually purplish, the teeth of the lip shorter than the galea.

3. **O. castilleioides**, Benth. At length diffuse and corymbosely branched; leaves from lanceolate to oblong, usually laciniate; the upper and the bracts cuneate-dilated and incisely cleft, green or the obtuse tips whitish or yellowish; spikes dense, short and thick: corolla nearly an inch long, dull white or purplish-tipped; lip ventricose-dilated.

SCROPHULARIACEÆ. (FIGWORT FAMILY.) 95

* * *Filaments pubescent; galea densely red-bearded; the obtuse tip incurved.*
4. **O. purpurascens**, Benth. Bracts and corolla usually crimson to rose-color. Distinguished by the bearded, hooked galea, and large stigma.

§ 2. TRIPHYSARIA, Benth.—*Lower lip of the corolla conspicuously 3-saccate, and very much larger than the slender galea, its teeth small, the tube filiform; stigma capitate, sometimes 2-lobed; bracts like the leaves and not colored.*

5. **O. pusillus**, Benth. Small and weak or diffuse, branched from the base, 3 or 4 inches high; leaves 1-2-pinnatifid, and bracts 3-5-parted into filiform divisions; flowers scattered, inconspicuous, shorter than the bracts; corolla purplish, 2 or 3 lines long; lip moderately 3-lobed; galea soon exposing the stamens.

6. **O. floribundus**, Benth. Slender, erect, 4 to 12 inches high; spike many-flowered, dense above; corolla white or cream-color, half an inch long; the tube twice the length of the calyx; stamens about the length of the soon open galea; the lip with 3 divergent oval sacs, their scarious teeth erect.

7. **O. erianthus**, Benth. Erect, a span or more high, much branched, pubescent; corolla sulphur-yellow, with the slightly falcate galea brown-purple; tube 6 to 8 lines long, filiform, densely pubescent, thrice the length of the calyx; the lip of 3 globular-inflated sacs, 1 to 2 lines long; the galea subulate, inclosing the stamens more strictly than the preceding.

Var. **roseus**, Gr. Corolla rose-purple, shorter.

8. **O. faucibarbatus**, Gr. Nearly smooth, less branched, and leaves with coarser divisions than the last; corolla with smaller sacs and less beard within the lip; the straight galea pale.

9. **O. lithospermoides**, Benth. Hirsute above; stem 4 to 12 inches high, strict, mostly simple, very leafy; bracts of the dense many-flowered spike about equaling the flowers; corolla an inch or less long, cream-color, often turning pale rose-color; sacs 3 lines deep; the teeth inconspicuous; anthers 2-celled.

11. CORDYLANTHUS. Nutt.

Calyx of an anterior and a posterior leaf-like division, or the former wanting. Corolla tubular, a little enlarging upward; the lips short and of nearly equal length; the lower very obtusely and crenulately 3-toothed; the upper straight and compressed, with the apex incurved. Style mostly hooked at the tip.—Branching annuals with alternate narrow leaves either entire or 3-5-parted; the floral ones not brightly colored. Flowers one to each bract, dull-colored, yellowish or purplish; the corolla not much exceeding the calyx.

§ 1. ADENOSTEGIA, Gr.—*Calyx 2-leaved; flowers short pedicelled or nearly sessile, subtended by 2 to 4 bractlets; floral leaves and bracts tipped with a gland.*

1. **C. filifolius**, Nutt. A foot or two high; leaves filiform; the lower entire, the

OROBANCHACEÆ. (BROOM-RAPE FAMILY.)

upper 3-5-parted, the floral with cuneate base and ciliate margins; corolla purplish, 6 to 9 lines long.

2. **C. pilosus**, Gr. Larger, soft-villous and hoary; the floral leaves 3-toothed at the tip; corolla yellowish with some purple, less than an inch long.

§ 2. HEMISTEGIA, Gr.—*Calyx 1-leaved; flowers without bractlets, each sessile in the axil of a claspiny bract; no glands at the tips of the leaves.*

3. **C. maritimus**, Nutt. Leaves smooth, somewhat fleshy, all entire; flowers in a capitate spike; corolla dull-purplish; pairs of filaments very unequal.—In salt marshes.

4. **C. mollis**, Gr. Stamens only 2, with smooth filaments; the upper leaves toothed or pinnatifid.—Salt marshes.

12. PEDICULARIS, Tourn.

Calyx 2-5-toothed, irregular. Corolla strongly bilabiate; the galea arched and laterally compressed; the lip 2-crested above, 3-lobed. Stamens 4, inclosed in the galea; anthers transverse, equally 2-celled.

1. **P. densiflora**, Benth. Nearly smooth, stout, becoming a foot or more high; leaves broad-lanceolate in outline, twice-pinnatifid or pinnately parted, and the divisions irregularly and sharply incised or toothed; the upper bracts of the dense elongated spike or raceme simpler; calyx-teeth, 5; corolla red or scarlet.

Order 44. OROBANCHACEÆ.

Root-parasitic herbs, destitute of leaves and green color. Distinguished from *Scrophulariaceæ* by the 1-celled ovary.

1. APHYLLON, Mitch.

Calyx 5-cleft, or 5-parted, regular or nearly so. Corolla tubular and curved, almost regular. or bilabiate. Stamens included; cells of the anthers deeply separated from below upward, mucronate at base. Stigma peltate or bilamellar.—Low pale or brownish herbs; the flowers yellowish or purplish.

* *Scapes or peduncles naked; corolla with an almost regular 5-lobed border.*

1. **A. uniflorum**, Gr. Corolla about an inch long, bluish purple, violet-scented.

2. **A. fasciculatum**, Gr. Scaly stem rising out of the ground 2 or 3 inches, bearing many peduncles; lobes of the calyx not longer than the tube; flowers dull yellow or purplish.

* *Stems rising above the ground; flowers bracteate; corolla plainly bilabiate.*

3. **A. comosum**, Gr. Low, branching at or near the surface of the ground; flowers

on slender pedicles in a corymb or short raceme; corolla rose-purple or purple, an inch or more long, or twice the length of the deeply parted calyx; anthers woolly.

4. **A. Californicum**, Gr. Flowers crowded in an oblong thyrsus or raceme; calyx lobes nearly equaling the tube of the yellowish or purplish corolla; anthers smooth or nearly so.

5. **A. tuberosum**, G. Flowers small, sessile in a compact cluster; yellowish.

Boschniakia, strobilacea, Gr., if found may be known by its resemblance to a spruce cone, 3 or 4 inches long, the flowers striped with white and brownish red; scale-like bracts brown.

Order 45. LABIATÆ.

Chiefly aromatic herbs with square stems, opposite simple leaves, and no stipules, bilabiate corolla, didynamous or diandrous stamens, and a 4-lobed ovary with a single style, forming seed-like nutlets in the bottom of the persistent calyx.—Flowers perfect, axillary. Calyx 3-5-toothed or cleft, or bilabiate. Stamens on the tube of the corolla. Style, 2-cleft at the apex; often unequally so, or one of the lobes obsolete; stigmas minute.

Tribe 1. SATUREIEÆ. Stamens erect or ascending; the posterior pair shorter or wanting; anthers 2-celled, and the short lobes never far separated, sometimes partly confluent but not blended. Upper lip of the corolla never hooded; all the lobes flat or nearly so.

* *The small corolla about equally 4-lobed; tube naked within.*

Stamens 4, nearly equal ....**Mentha.** 1
Stamens 2, with anthers; posterior pair sterile or wanting ....**Lycopus.** 2

* * *Corolla bilabiate; no hairy ring within the base of the tube.*

+ *Calyx about equally 5-toothed and 13-nerved; style beardless.*

Flowers glomerate-capitate. Stamens 4, straight.
Stamens distant and divergent.................................**Pycnanthemum.** 3
Stamens exserted ....**Monardella.** 4

Flowers solitary or clustered in the axils.
Stamens 4, curving, shorter than the corolla........................**Micromeria,** 5

+ + *Calyx unequally and deeply 5-cleft, mostly 13-nerved; style bearded above.*
Stamens 4, sometimes the upper pair sterile.........................**Pogogyne.** 6

* * * *Corolla not manifestly bilabiate; a hairy ring at the base of the tube within.*
Shrubby. Flowers large, campanulate..................................**Sphacele.** 7

Tribe 2. MONARDEÆ. Stamens only 2, fertile, the upper pair rudimentary or wanting; anthers apparently or really of a single linear-oblong cell, or of 2 cells widely separated upon the ends of a filament-like connective.

LABIATÆ. (MINT FAMILY.)

Connective longer than the filament itself, which it strides, a narrow anther-cell at its upper end, a smaller one or a long process at the lower. . . . . Salvia. 8
Connective much shorter than the slender filament and continuous or barely articulated with its apex, or apparently none; anther 1-celled, no rudiment of the second cell below . . . . Audibertia. 9

Tribe 3. STACHYDEÆ. Stamens 4, with anthers, ascending and parallel under the concave or galeate upper lip of the corolla. Calyx 5-10-nerved. Herbage less aromatic than the preceding tribes.
Calyx with a projection on the upper side, casque-shaped . . . . Scutellaria. 10
Calyx bilabiate. Filaments 2-forked, one fork bearing the anther . . . . Brunella. 11
Calyx 5-10-nerved, nearly equally 5-toothed . . . . Stachys. 12

Tribe 4. AJUGOIDEÆ. Stamens parallel, and protruding from the cleft on the upper side of the corolla; the anterior longer.
Corolla with 5 similar oblong lobes . . . . Trichostema. 13

1. MENTHA, L. MINT.

Calyx about equally 5-toothed. Corolla with a short included tube, and a campanulate border; the upper lobe broadest, entire or emarginate. Odorous herbs, with very small flowers in dense clusters forming an apparent whorl in the axils or spikate at the tops of the branches.

1. M. Canadensis, L. Leaves from oblong-ovate to almost lanceolate, sharply serrate, acute, short-petioled; flowers all in axillary clusters, whitish or purplish.

2. LYCOPUS, Tourn. WATER HOREHOUND.

Chiefly distinguished from Mèntha by the stamens. Flowers white, in false whorls.
1. L. lucidus, Turcz., var. Americanus, Gr. The subterranean runners producing tubers; leaves lanceolate, 2 to 4 inches long, coarsely serrate, sessile or nearly so.

3. PYCNANTHEMUM, Michx.

Corolla short, with tube hardly exceeding the calyx. Anther-cells close and parallel. Perennial erect herbs with small flowers.
1. P. Californicum, Torr. About 2 feet high, corymbosely branched, sweet-odorous, whitened with soft pubescence, or in age smoothish: leaves from ovate to ovate-lanceolate, closely sessile by a slightly cordate or roundish base, sparingly denticulate or entire; heads of flowers very dense at the summit, white-villous; flowers whitish.

4. MONARDELLA, Benth.

Marked by the flowers compacted in terminal heads involucrate with bracts, flesh-color or purple.

LABIATÆ. (MINT FAMILY.)

* *Perennial, in tufts from a procumbent and almost woody base.*

1. **M. villosa**, Benth. Soft-pubescent or villous a foot or two high; leaves ovate, often with a few obtuse teeth, being 6 to 10 lines long, petioled. Sometimes nearly smooth.

** *Annual; leaves entire or undulate.*

2. **M. undulata**, Benth. A span to a foot or more high; leaves from oblong spatulate to nearly linear with a narrowed base, obtuse, undulate-margined, about an inch long; bracts and calyx villous; corolla rose-color. Has the odor of Peppermint.

3. **M. Breweri**, Gr. A span or more high; leaves oblong or ovate, pinnately veined, the larger an inch long; bracts broadly ovate, cuspidate, whitish-scarious, the outer pinnately and the inner nervosely 7-9-ribbed; corolla rose-purple.

4. **M. Douglasii**, Benth. Loosely branched; leaves lanceolate, an inch long, tapering into the petiole; the silvery white or purple-tinged bracts mostly transparent, with a strong marginal vein connected with the midrib by pinnate veins.—Strong-scented; corolla deep rose-color.

5. MICROMERIA, Benth.

Calyx not gibbous. Corolla short; upper lip erect, flattish, entire or emarginate; lower spreading, 3-parted.—Low plants, sweet-odorous, with small axillary flowers.

1. **M. Douglasii**, Benth. YERBA BUENA. Perennial herb, with long slender creeping and trailing stems; leaves round-ovate, thin, sparingly toothed, short petioled, an inch long or less; flowers mostly solitary on a long filiform 2-bracteolate peduncle; corolla purplish or white, 4 lines long.

2. **M. purpurea**, Gr. Erect, much branched; leaves lanceolate, acuminate, sparsely serrate; flowers in umbel-like clusters; corolla purple-blue, 2 lines long.

6. POGOGYNE, Benth.

Calyx cleft to below the middle; the 2 lower teeth longer; corolla straight, tubular-funnelform, with short lips; the erect and entire upper lip and the three lobes of the spreading lower one oval and somewhat alike. Stamens with the upper shorter pair sometimes sterile; the anther cells parallel and pointless. Style somewhat exserted, bearded above.—Low annuals, sweet-aromatic; with oblong or oblanceolated leaves narrowed into a petiole; flowers mostly crowded and interrupted spicate; bracts and calyx hirsute-ciliate; the corolla blue or purplish.

* *Stamens all four with anthers; style conspicuously bearded above, and its subulate lobes almost equal; corolla 6 to 9 lines long; flowers densely crowded into an oblong cylindrical spike, which is conspicuously white-hirsute with the long, stiff, ciliate hairs of the calyx.*

1. **P. Douglasii**, Benth. Rather stout, a span to a foot high; leaves veiny, some-

times sparingly toothed; bracts linear, acute; lower lobes of the calyx much longer than the others.

2. **P. parviflora**, Benth. Smaller; bracts mostly obtuse; corolla 5 or 6 lines long.

* * *Upper stamens sterile; style sparingly hairy, its lobes very unequal; flowers barely 2 lines long.*

3. **P. serpylloides**, Gr. Stems 3 to 6 inches high; leaves obovate-oval or spatulate, 2 or 3 lines long; lower flowers remote and often solitary; the upper usually interruptedly spicate.

7. SPHACELE, Benth.

Calyx thin, membranaceous and reticulated. Corolla with 5 broad, rather erect lobes, the lower one longest. Anther cells diverging. Somewhat shrubby, veiny-leaved.

S. calycina, Benth. Villous-pubescent or tomentose, leafy, 2 to 5 ft. high; leaves 2 to 4 inches long, ovate or oblong crenate or serrate, or almost entire; the floral, ovate-lanceolate, sessile; flowers an inch long, mostly solitary in the upper axils, purplish or lead-color.

8. SALVIA, L. SAGE.

Calyx bilabiate. Corolla deeply 2-lipped, the upper lip erect, straight or falcate, 2-lobed, the lower spreading or drooping, its middle lobe sometimes notched or obcordate. In our species the upper lip of the calyx is longer than the lower, 3-2-toothed; the lower 2-parted; the teeth spinulose; corolla ringent.

1. **S. carduacea**, Benth. White-woolly with cobwebby hairs; stems nearly naked, surrounded at the base with thistle-like leaves; head-like false whorls 1 to 4, an inch or more in diameter, about equaling the involucre of spiny-toothed bracts; corolla 10 to 12 lines long, blue or purple.

2. **S. Columbariæ**, Benth. (CHIA.) Soft pubescent; flower whorls 1 or 2; involucrate bracts, sometimes purplish; corolla 3 or 4 lines long, blue; leaves not spinescent.

9. AUDIBERTIA, Benth.

Sufficiently distinguished from *Salvia* in the synopsis.—Mostly hoary perennials, herbaceous or shrubby; with rugose-veiny, crenulate, sage-like leaves, and densely capitate-glomerate flowers.

1. **A. grandiflora**, Benth. Stems 1 to 3 feet high from a somewhat woody base; lower leaves 3 to 8 inches long; floral ones broadly ovate and membranaceous; corolla an inch and a half long; purple-crimson; stamens much exserted.

2. **A. humilis**, Benth. A span high, cespitose; leaves mainly radical; spike of 3 or 4 small, sessile, head-like clusters; corolla half an inch long or less, bluish purple.

3. **A. stachyoides**, Benth. Shrubby, 3 to 8 feet high; style and stamens little exserted; corolla about as the last.

10. SCUTELLARIA, L. Skullcap.

Calyx, with two entire lips and a gibbous projection on the back, closed after flowering. Corolla, with an elongated and curved ascending tube, a dilated throat, an erect arched or galeate upper lip, with which the lateral lobes appear to be connected; the anterior lobe appearing to form the whole lower lip.—Herbs, not aromatic; with single axillary, rather conspicuous flowers.

1. **S. angustifolia**, Pursh. A span to a foot high; leaves about an inch long; the radical ones often roundish or even cordate; corolla blue or violet, an inch long, with a slender tube; lower lobe villous inside.—Ours is mainly

Var. **canescens**, Gr. A form with soft, hoary pubescence, and the tube of the corolla bent so as to throw the upper part backward.

2. **S. Californica**, Gr. Puberulent; stems 8 to 20 inches high, slender; leaves from lanceolate-oblong to oval-ovate; the lower an inch or more long, often serrate; upper gradually reduced to half an inch or less; lips of the yellowish corolla about equal.

3. **S. tuberosa**, Benth. Soft, pubescent or villous; stems slender, erect and short, or trailing a foot in length; the filiform subterranean shoots bearing tubers; leaves mostly ovate, coarsely and obtusely few-toothed or entire, 5 to 18 lines long; corolla deep blue or violet.

11. BRUNELLA, Tourn. Self-heal.

Calyx-lips closed in fruit. Corolla with ascending tube, open lips, and slightly-contracted orifice; upper lip arched and entire; lower 3-lobed, its middle lobe drooping, rounded, concave, denticulate.—Low perennials, the flowers crowded in a terminal oblong or cylindraceous head or spike.

1. **B. vulgaris**, L. A span to a foot or more in height; leaves ovate or oblong, slender-petioled; corolla violet, purple, or rarely white; calyx purplish.

12. STACHYS, L.

Corolla with cylindrical tube not dilated at the throat; the upper lip erect and concave or arched; the lower spreading, its middle lobe larger. Stamens ascending under the upper lip; filaments naked; anthers approximate in pairs, 2-celled.—Herbs, not aromatic, with flowers clustered, capitate, or scattered, often spicate at the end of the branches; flowers sessile or nearly so.

* *Corolla white or whitish; the upper lip bearded or woolly on the back; herbage tomentose or soft hairy.*

1. **S. ajugoides**, Benth. A span to a foot high; silky-villous with whitish hairs; leaves oblong, very obtuse, crenately serrate, 1 to 3 inches long, the upper sessile; flowers about 8 in the axils of the distant upper leaves, and loosely leafy-spicate at the summit.—Moist ground.

VERBENACEÆ. (VERVAIN FAMILY.)

2. **S. albens**, Gr. Soft-tomentose with whitish wool, 3 to 5 ft. high; leaves mostly cordate at base, obtuse, crenate, 2 or 3 inches long; flowers several or many in capitate clusters which usually exceed the small floral leaves and form an interrupted spike; corolla white with purple dots on the lower lip.

3. **S. pycnantha**, Benth. Very hirsute, with long and mostly soft spreading hairs, not white, two feet high or more; flowers in a dense cylindraceous naked spike (an inch or two long), exceeding the small bract-like floral leaves except in the lowest and sometimes rather distant clusters; corolla white or cream-color, with purple on the lower lip. (?)

* * *Corolla purple, the upper lip hairy on the back; pubescence somewhat hispid; no tomentum.*

4. **S. bullata**, Benth. Stem retrorsely hispid, especially on the angles, 1 to 3 ft. high; leaves somewhat rugose, nearly all petioled, 1 to 2 inches long; flowers usually 6 in the false whorls, these rather distant, forming a narrow interrupted spike; lower lip of the corolla fully as long as the tube, 4 or 5 lines long, the upper half as long.—Variable.

* * * *Tube of the rose-red corolla twice as long as the calyx, 6 to 9 lines long.*

5. **S. Chamissonis**, Benth. Stem 2 to 5 ft. high, stout, mostly rough-hispid, with retrorse rigid bristles; leaves 2 to 5 inches long; lips of the corolla pubescent outside.—Wet ground.

13. TRICHOSTEMA, L. BLUE-CURLS.

Calyx campanulate and almost equally 5-cleft. Corolla with short or slender tube and an almost equally 5-parted limb. Stamens with long capillary curved filaments, sometimes cohering at the base.—Strong scented herbs; with entire leaves, and blue or purple corolla and stamens. In ours the flowers are in cymose axillary clusters, somewhat raceme-like in age; the corolla about 3 lines long, and the stamens twice as long or more.

1. **T. laxum**, Gr. Minutely soft pubescent, about a foot high, simple or loosely branched from the base; leaves rather distant, lanceolate or oblong-lanceolate, tapering into a petiole at the base; flower clusters distinctly peduncled, usually forked and in age equaling the leaves; corolla almost smooth.

2. **T. lanceolatum**, Benth. Leafy; leaves much longer than the internodes, lanceolate or ovate-lanceolate, sessile by a broad base, 3–5-nerved, an inch or less long; flower clusters nearly sessile, short, one-sided; corolla somewhat pubescent.—Its odor sickening, tarry.

Order 46. VERBENACEÆ.

Herbs or shrubs differing from *Labiatæ* mainly in the ovary and fruit, which is undivided and 2–4-celled, at maturity either dry and splitting into as many 1-seeded nutlets, or drupaceous, containing as many little stones.

PLANTAGINACEÆ. (PLANTAIN FAMILY.)

1. VERBENA, L.

Calyx 5-toothed, one tooth often shorter. Corolla salver-form, the limb unequally 5-cleft. Stamens included, the upper pair sometimes sterile. Stigma unequally lobed. Ovary 4-celled.—Herbs with small flowers, ours about 2 lines in diameter.

1. **V. officinalis**, L. Some of the lower leaves pinnatifid; spikes mostly solitary, filiform; corolla purple or lilac, 2 or more lines in diameter.
2. **V. hastata**, L. Stouter and taller, 3 to 6 ft. high; leaves serrate or incised, the lower hastate-3-lobed; spikes panicled, densely flowered; corolla blue, 2 lines in diameter.
3. **V. prostrata**, R. Br. Soft hirsute, diffuse, a foot high; villous spikes long; corolla violet or blue.

Order 47. PLANTAGINACEÆ.

Stemless herbs with flowers in spikes, the 4-cleft regular corolla dry and scarious.

1. PLANTAGO, L. Plantain.

Flowers in spikes or heads, bracteate. Calyx of 4 persistent sepals free from the ovary. Stamens 2 or 4 on the corolla alternate with its lobes, anthers versatile. Style filiform, bearded above.—Stemless herbs with nerved or ribbed radical leaves and naked scapes of small greenish flowers.

* *Flowers with 4 stamens.*

1. **P. major**, L. Mostly smooth; leaves ovate or broadly oblong, abruptly contracted into a channeled petiole, 5–7-ribbed; spike long and slender; capsule 7–16-seeded.
2. **P. hirtella**, HBK. Leaves smooth, rather fleshy, oblanceolate to obovate, 3–7-ribbed, tapering into a narrow base or wing-margined petiole; scape 1 to 3 ft. high; flowers large.
3. **P. lanceolata**, L. Mostly hairy; leaves lanceolate, 3–5-ribbed; scape deeply grooved.
4. **P. maritima**, L. Leaves linear, fleshy; scapes usually short.
5. **P. Patagonica**, Jacq. Leaves linear to filiform, thin, usually silky-woolly.—Dry ground.

* * *Flowers with 2 stamens.*

6. **P. Bigelovii**. Gr. Leaves linear; small.—Salt marshes.

INDEX.

	PAGE		PAGE		PAGE
Acœna	54	Capsella	24	Erysimum	23
Acer	37	Cardamine	23	Erythræa	75
Aconitum	19	CARYOPHYLLACEÆ	27	Eschscholtzia	21
Adenostoma	63	Castilleia	93	Eucharidium	62
Æsculus	37	Ceanothus	36	Euonymus	35
Alchemilla	54	CELASTRACEÆ	35		
Amelanchier	64	Cephalanthus	65	Ficoideœ	63
Amorpha	47	Cerastium	28	Fragaria	53
Amsinckia	84	Cercocarpus	52	Fraxinus	73
Anacardiaceœ	38	Cheiranthus	23	FUMARIACEÆ	22
Anagallis	73	Chimaphila	71		
Anemone	16	Circœa	62	Galium	65
Antirrhinum	90	CISTACEÆ	25	Garrya	64
Aphyllon	96	Clarkia	61	Gaultheria	70
Apocynaceœ	73	Claytonia	30	Gentiana	75
Apocynum	73	Clematis	16	GENTIANACEÆ	74
Aquilegia	18	Collinsia	90	GERANIACEÆ	33
Arabis	23	Collomia	76	Geranium	33
Aralia	63	COMPOSITÆ	66	Gilia	76
Araliaceœ	63	CONVOLVULACEÆ	86	Githopsis	67
Arbutus	69	Convolvulus	86	Glaux	73
Arctostaphylos	69	Cordylanthus	95	Glycyrrhiza	47
Arenaria	28	CORNACEÆ	63	Godetia	61
Armeria	72	Cornus	63	Gomphocarpus	74
ASCLEPIADACEÆ	73	Cotyledon	58		
Asclepias	73	CRASSULACEÆ	58	Helianthemum	25
Astragalus	47	Cressa	87	Heliotropium	84
Audibertia	100	CRUCIFERÆ	22	Heterocodon	68
Azalea	70	Cucurbitaceœ	63	Heteromeles	54
		Cuscuta	87	Heuchera	57
BERBERIDACEÆ	19	Cynoglossum	86	Horkelia	53
Berberis	19			Hosackia	44
Boisduvalia	62	Datura	88	HYDROPHYLLACEÆ	80
BORRAGINACEÆ	83	Delphinium	18	HYPERICACEÆ	30
Boschniakia	97	Dendromecon	21	Hypericum	30
Boykinia	56	Dicentra	22		
Brasenia	20	Dodecatheon	72	Jussiæa	59
Brassica	23	Downingia	67		
Brunella	101			LABIATÆ	97
		Ellisia	81	Lathyrus	49
Calandrinia	29	Emmenanthe	83	Lavatera	31
Calycanthaceœ	55	Epilobium	59	LEGUMINOSÆ	38
Calycanthus	55	ERICACEÆ	68	Lepidium	24
Campanula	68	Eriodictyon	83	Lepigonum	29
CAMPANULACEÆ	67	Eritrichium	85	Lewisia	30
CAPRIFOLIACEÆ	64	Erodium	33	Limosella	93

INDEX.

	PAGE
Limnanthes	34
LINACEÆ	32
Linaria	90
Linum	32
LOASACEÆ	62
Lobeliaceæ	67
Lonicera	65
Lupinus	39
Lycopus	98
Lythraceæ	59
Lytarum	59
Malva	31
MALVACEÆ	31
Meconopsis	21
Medicago	44
Megarrhiza	63
Melilotus	43
Mentha	98
Mentzelia	62
Menyanthes	75
Mesembryanthemum	63
Microcala	75
Micromera	99
Mimulus	92
Mollugo	63
Monardella	98
Myosurus	17
Negundo	38
Neillia	51
Nemophila	81
Nicotiana	89
Nuphar	20
Nuttalia	51
NYMPHÆACEÆ	20
Œnothera	60
Oleaceæ	73
ONAGRACEÆ	59
OROBANCHACEÆ	96
Orthocarpus	94
Oxalis	34
Pæonia	19
PAPAVERACEÆ	20
Peclocarya	86

	PAGE
Pedicularis	96
Pentstemon	91
Phacelia	81
Philadelphus	57
Photinia	54
Pickeringia	39
Pirus	54
PLANTAGINACEÆ	103
Plantago	103
Platystemon	20
Platystigma	21
Plectritis	66
PLUMBAGINACEÆ	71
Pogogyne	99
POLEMONIACEÆ	75
Polemonium	80
Polygala	27
Polyga'aceæ	27
PORTULACACEÆ	29
Potentilla	53
PRIMULACEÆ	72
Prunus	50
Psoralea	46
Ptelea	34
Pterospora	71
Pycnanthemum	98
Pyrola	71
Raphanus	25
RANUNCULACEÆ	16
Ranunculus	17
RHAMNACEÆ	35
Rhamnus	35
Rhododendron	70
Rhus	38
Ribes	57
Romanzoffia	83
ROSACEÆ	49
Rosa	54
RUBIACEÆ	63
Rubus	52
RUTACEÆ	34
Salvia	100
Sambucus	64
SAPINDACEÆ	37
Sarcodes	71

	PAGE
SAXIFRAGACEÆ	55
Saxifraga	55
SCROPHULARIACEÆ	89
Scrophularia	91
Scutellaria	100
Sedum	58
Sidalcea	31
Silene	27
SOLANACEÆ	88
Solanum	88
Specularia	67
Spergula	29
Spiræa	51
Sphacele	100
Stachys	101
Statice	72
Stellaria	28
Symphoricarpus	64
Tellima	56
Thalictrum	17
Thermopsis	39
Thysanocarpus	25
Tiarella	57
Tillæa	59
Trichostema	102
Trientalis	73
Trifolium	42
Tropidocarpum	24
UMBELLIFERÆ	63
Vaccinium	69
VALERIANACEÆ	66
Vancouveria	20
VERBENACEÆ	102
Verbena	103
Veronica	93
Vicia	48
VIOLACEÆ	25
Viola	26
Vitaceæ	37
Vitis	37
Whipplea	57
Zauschneria	59

ERRATA.

In the Preface, for *Edogenous* read *Endogenous*.
Page 4, under **Lewisia**, for *L. rediva* read *L. rediviva*.
Page 21, under Eschscholtzia, for *torus* read *torus*.
Page 23, for *acumbent* read *accumbent*.
Page 25, under **Raphanus**, for *Raphinistrum* read *Raphanistrum*.
Page 29, under **Lepigonum**, for *pediceled* read *pedicelled*.
Page 30, under **Claytonia**, for *C. Sibrica* read *C. Sibirica*.
Page 45, under **Hosackia**, for *H. Trachycarpa* read *H. brachycarpa*.
Page 50, under **Rosaceæ**, for *Neileia* read *Neillia*.
Page 57, under **Philadelphus**, for *P. Gordianus* read *Gordonianus*.
Page 59 62, under **Onagraceæ**, for **Eucharideum** read **Eucharidium**.
Page 74, under **Asclepias**, for *A. restitia* read *A. restita*.
Page 80, under **Hydrophyllaceæ**, for **Emenanthe** read **Emmenanthe**.
Page 97, under **Tribe 1**, for *part* read *pair*.

In several places *setaceously* and *stipitate* are incorrectly spelled; narrowly is wanting an *r* on page 41, and a few hyphens are omitted in compound adjectives.

www.ingramcontent.com/pod-product-compliance
Lightning Source LLC
Chambersburg PA
CBHW020149170426
43199CB00010B/954